Contents

Contents

List of Illustrations

The plate section appears between pp. xvi and 1.

Preface

Domes have been used as roofs since ancient times, but their use has become increasingly popular in recent years, for example as roofs to libraries, mosques, synagogues, and water, sewage and chemical tanks. This book is principally concerned with reinforced concrete and prestressed concrete domes. The elastic analysis used for designing these domes in structural concrete is also relevant to the design of domes in laminated timber, fibreglass and masonry.

Membrane theory considers the loading to be carried by only membrane (in planar) forces (i.e. analogously to a soap bubble) and not bending moments. For a dome consisting of a hemisphere, and carrying a uniformly distributed load, the boundary forces that are needed for equilibrium are vertical and these may be provided, for example, by supporting the dome with masonry walls. But for a similarly loaded dome comprising a less than hemispherical portion of a sphere, the boundary forces are inclined. If the periphery is supported vertically by a masonry wall and horizontally, to avoid spreading, by a reinforced concrete ring beam then this beam will stretch (strain) under the ring tension it carries. This movement is generally of a different amount to the strain in the periphery of the dome due to membrane forces. So if one considers not only equilibrium of forces (e.g. as would be done by Hillerborg's strip and advanced methods for slabs) but the elastic matching up of displacements of points of contact of ring beam and dome shell periphery, then transverse moments and corresponding radial shears need to be taken into account, as well as the membrane forces in the dome shell.

This kind of elastic design, satisfying boundary conditions between shell and edge beam, is very important for designing domes, in order to control cracks reasonably well and to assess edge displacements. Although shells have commonly been designed in this way since about 1950, occasionally the equilibrium of forces only method is still used for domes. This is probably because certain past and present design books and handbooks present this method for dome designers who have no previous knowledge of shell design or analysis. These particular publications deal mainly with non-shell design and have a very small section on domes, a situation which can be misleading to some designers.

The elastic method of design (approved for domes by ACI 318–89) matching up boundary conditions as described above, is much more complex and arduous for the designer than the equilibrium of forces only method. The author therefore decided to program the analysis for a microcomputer and so produced the design tables given in this book. These can be used by a designer with minimum time and effort.

The author commenced by programming formulae from one book on the design of domes. These solved the example in that book. He then applied the program to examples from another two books, but obtained wildly differing results—the problem being that the books give examples with limitations which are presumably made to simplify the calculations for the example but can be misleading to the designer.

The programming exercise explored all the problems and errors in various previous books. The present book, therefore, aims to present reliable formulae for use in design and useful tables for designers of practical domes. *Concrete dome roofs* is the first book of its kind to be published in nearly two decades.

Clients usually wish to make full use of their land and do not choose the spans of their domes rounded-off to the nearest metre simply to suit design tables. There is the problem of SI, British Imperial and traditional metric units, where the kilogram is used for both force and mass and the metre, centimetre and millimetre (dependent on country) are used for length (mainly mainland Europeans still think in and use these). Therefore, the tables have been devised close enough together to allow interpolation for practical designs. Examples show how to use SI units based on metres (length), newtons (force), kg. (mass) and seconds as well as those based upon inches, feet (length), pounds, kips (force and mass), and seconds.

For the benefit of those who still mainly think in British Imperial/USA units, these are often given as well as SI units in the text. One design is led with British/USA units, giving SI equivalents throughout. The designer can now, in the author's opinion, with the aid of this book, reliably and rapidly design a large range of practical domes.

The designer is also able, with the help of this book, to make drawings of domes, prepare a Bill of Quantities and, from this, estimate the cost of construction. In this connection the cost of curved shuttering seems to have always been acceptable in view of the greatly reduced quantities of concrete, reinforcement and shuttering compared to beam and slab construction, which would also involve many internal columns and their foundations (bases or piles for example). Domes have sometimes been used over service reservoirs in order not to puncture the floor slab and to avoid the cost of the columns and particularly the foundations when piling has been necessary.

If the dome contract is won the designer may quickly and easily use the design tables and examples in this book to design and detail the required structures.

A circular tank may be basically more economical than a rectangular tank, and can be prestressed advantageously. Such tanks are particularly suited to dome roofing. Quoting from Faber and Mead (Ref. 2.6) 'the formwork for domes is relatively expensive but thickness and weight of concrete and reinforcement can be kept small even for large spans ... lightness of shells enables cheaper substructures and foundations ... domes are most commonly of spherical form'. They also report that 'the first shell concrete roofs were domes, constructed mainly in Germany and used for covering large public spaces, the Market Hall at Leipzig, 1929, is a typical example'.

Designs based on British codes (the ones best known to the author) using elastic theory have a general record of success with reinforced, prestressed and shell concrete over at least 40 years. The author has been called in to investigate many kinds of problems and has never found faults in a structure designed properly by these methods (except in the case of the shear strength of plain webbed members—see various publications by the author).

Apart from the usual library literature searches, an online search was made of the Compendex database using the DIALOG system based in Palo Alto, California, USA.

Elastic Design Appendices

A set of 240 tables (all worked out from first principles, with equations rigorously checked) based on research and information gathered from world-wide sources are presented in useful appendices that comprise the greater part of the book. Appendix 5 gives data for non-prestressed domes, and Appendix 6 data for prestressed domes; both appendices provide analyses for standard shell domes, domes with a ring beam acting as a rain-water gutter, and domes with a ring beam that has a clean soffit (visually comparable to a shell dome). Assuming little familiarity with BS 8110 and ACI 318–89, and with elastic and shell analyses, the tables will help the designer to obtain by hand the often complex results of computer analysis and output for the elastic method of design of domes, and will enable the engineer to design with speed and economy. The tables take into account limitations that may not be specifically drawn to the attention of the designer that would in fact invalidate the more general practical designs required.

The domes analysed in these appendices have edges (springing lines) at only 28 degrees on either side of the verticals through their crowns. They thus have a self-weight heavier than any suction that can occur in the British Isles, and internationally, due to hurricanes. The design tables of this book can, therefore (according to CP3, BSI) be used without modification for situations where hurricanes are or may be experienced. The domes shown in the frontispiece (located in Florida, USA) are generally lighter in weight than the ones given in the appendices, since the author has been more conservative with the design recommendations regarding thin shell buckling.

Other Appendices

The book also includes appendices covering:

- Wind pressure on domes
- Snow loading on domes
- Unit conversion

Units

Both SI and Imperial units are given throughout the worked examples in the book; and the design tables of Appendices 5 and 6, though given in SI units, have dome plan diameters that vary in such small increments that will allow interpolation for all practical designs in any unit. Reference has also been made to both British and American codes of practice.

The Author

C B Wilby is Emeritus Professor of Engineering at the University of Bradford. He has taught shell, folded plate and prestressed concrete design and analysis for many years, and has published extensively in these areas. He has long experience of the design, analysis, detailing, construction, site and laboratory testing, research and rectification of existing shells, as well as long experience of prestresssed concrete.

He was an early designer of shell roofs and of prestressed concrete in the British Isles and has subsequently been involved in the design and construction of shell roofs and research into these structures and allied problems over many years, and has now produced the tables in this book which practical designers may wish to consult, based upon this considerable experience. He has previously published three similar books of design tables and graphs for designing cylindrical and conoidal shell roofs.

C B Wilby
Harrogate 1993

Disclaimer of Warranty

Acknowledgements

The author is indebted to the most useful help kindly and pleasantly given by Anne Costigan of the Bradford University library; to Chris B Wilby for considerable help in the transfer of programs from the HP85 to, and the use of, the IBM clone computer; and to Mark Stainburn Wilby for some computing help.

The author and publishers are indebted to the following for permission to reproduce copyright material:

Theodore R Crom, P E, Chairman, The Crom Corporation. Florida, USA.

Nomenclature

NB See Appendix 4 for symbols used in the design tables of Appendices 5 and 6.

a	radius of shell. NB $(a - h/2)$ is radius of soffit (undersurface) of shell, see Figs 1.1, 1.2
b	breadth of ring beam, see Fig. 1.1
C	see Eq. 2.1
$(C1)$	see Eq. 1.4
$(C2)$	see Eq. 1.5
$(C3)$	see Eq. 1.10
$(C4)$	see Eq. 1.11
$(C5)$	see Eq. 1.16
$(C6)$	see Eq. 1.17
$(C7)$	see Eq. 1.21
$(C8)$	see Eq. 1.28
$(C9)$	see Eq. 1.29
d	depth of ring beam, see Fig. 1.1
e	distance from centroid of ring beam to the tangent to the shell of radius a at the springing line, see Figs 1.3, 1.5
E_c	modulus of elasticity (Young's modulus) of concrete
f_b	fb in Appendix 4; stress in bottom fibre of ring beam, positive denotes tensile stress
f_t	ft in Appendix 4; stress in top fibre of ring beam, positive denotes tensile stress
$(F1)$	see Eq. 1.12
$(F2)$	see Eq. 1.20
$(F3)$	see Eq. 1.24
$(F4)$	see Eq. 1.30
h	equivalent constant thickness, see Eq. 2.5
h_1	thickness of shell where not thickened, see Fig. 1.1
h_2	maximum thickness of thickening, see Fig. 1.1
h_a	average thickness of thickened region
H	internal horizontal force necessary to satisfy boundary conditions between shell and ring beam, see Fig. 1.3, Section 1.2

H_r	horizontal force on ring beam, see Eq. 1.45
M	internal bending moment necessary to satisfy boundary conditions between shell and ring beam, see Fig. 1.3 and Section 1.2
M_r	couple about centroid of ring beam, see Eq. 1.46
M_{rx}	bending moment about horizontal axis of ring beam, see Eq. 1.48
M_ϕ, M_ϕ'	see Fig. 1.4
$N_\phi, (N1)$	inplanar shell force per unit length, see Fig. 1.4
$N_\theta, (N2)$	inplanar shell force per unit length, see Fig. 1.4
q	loading per unit of curved surface area of shell, see Section 1.3.1
Q_ϕ, Q_ϕ'	radial shear force per unit length, see Fig. 2.6
r	half of plan diameter or span of shell part of dome, see Fig. 1.2
S	curved distance from springing to end of thickening
T_r	axial tension in ring beam, see Eq. 1.47
V_r	vertical force on ring beam, see Eq. 1.44
y	horizontal outward deflection of a point on the springing line, see Fig. 1.3
y_0	height of springing line (where shell and ring beam meet) above horizontal central axis of ring beam, see Fig. 1.2
$(Y1)$	see Eq. 1.8
$(Y2)$	$(C5)$
$(Y3)$	see Eq. 1.23
$(Y4)$	see Eq. 1.27
$(Y5)$	see Eq. 1.52
α	angle between crown and springing line of shell, see Fig. 1.2
$\lambda, (L)$	see Eq. 1.6
ϕ	angle in vertical plane from the crown, see Fig. 1.4
ψ	angle of rotation of shell edge and ring beam at line of contact, see Fig. 1.3
θ	angle in horizontal plane, see Fig. 1.4
$\nu, (V)$	Poisson's ratio

The Greek Alphabet

1.	A	α	alpha		9.	I	ι	iota
2.	B	β	beta		10.	K	κ	kappa
3.	Γ	γ	gamma		11.	Λ	λ	lambda
4.	Δ	δ	delta		12.	M	μ	mu
5.	E	ε	epsilon		13.	N	ν	nu
6.	Z	ζ	zeta		14.	Ξ	ξ	xi
7.	H	η	eta		15.	O	o	omicron
8.	Θ	θ	theta		16.	Π	π	pi

17.	P	ρ	rho
18.	Σ	σ	sigma
19.	T	τ	tau
20.	Y	υ	upsilon

21.	Φ	ϕ	phi
22.	X	χ	chi
23.	Ψ	ψ	psi
24.	Ω	ω	omega

Other symbols used in mathematics:

∇ del

∂ partial differential

Plate 1 Mosque, Tripoli (courtesy of former C & CA, an Esso photograph).

Plate 2 Sports stadia, Rome (courtesy of former C & CA).

Plate 3 University, Jerusalem (courtesy of former C & CA).

Plate 4 Concrete domes (courtesy of former C & CA).

Plate 5 Chemical-waste disposal tank with dome cover, UK.

Plate 6 Water storage tank, Florida. Dome 48.77 m (160 ft) span (courtesy of The Crom Corporation, Prestressed Composite Tanks, 250 SW 36th Terrace, Gainsville, Florida 32607, USA).

Plate 7 Water storage tank, Florida. Dome 47.24 m (155 ft) span (courtesy of The Crom Corporation, Florida, USA).

Plate 8 Water tower, Florida. Dome 14 m (46 ft) span (courtesy of The Crom Corporation, Florida, USA).

Plate 9 Water storage tank, Kentucky, USA. Dome 39.62 m (130 ft) span (courtesy of The Crom Corporation, Florida, USA).

Plate 10 Water storage tank, Georgia, USA. Dome 13.7 m (45 ft) span (courtesy of The Crom Corporation, Florida, USA).

Plate 11 Formwork/scaffold to tank and covering dome, USA (courtesy of The Crom Corporation, Florida, USA).

1 Survey of Design Formulae and Examples

1.1 Introduction

Many publications[1.1−1.8] contain design formulae for the construction of domes, and occasionally contain examples to assist the practical designer. Whilst these are of benefit, it must be emphasized that caution should be exercised when applying any example to a given situation; for, in the real world, there may be limitations that the examples do not take into account, and that would result in inadequacies in the construction. The aim of this book has, therefore, been to present a series of *reliable* design tables (Appendices 5 and 6), based on reliable formulae, that the engineer with little familiarity of design criteria can refer to for practical constructions without recourse to arduous analysis and computer calculations.

During the compilation of these tables the opportunity has been taken to rederive the basic mathematical formulae, and to check these against publications in Chapter 1.

1.2 Practical Dome Details

Chapter 1 looks at design formulae obtained from an elastic analysis of a practical spherical dome (Fig. 1.1) under non-prestressed (Section 1.3) and prestressed (Section 1.5) conditions. Fig. 1.1 shows, for a practical spherical dome, that the top of the ring, or edge, beam can be higher than the dome springing point so that a rain-water gutter can be formed between the dome shell and upstand edge beam as is commonly done in practice for many types of shells.[1.8−1.11] Fig. 1.2 gives dimensions for Fig. 1.1. Fig. 1.3 shows (a) the ring beam and (b) the dome shell of Fig. 1.2, and the indeterminate horizontal force per unit length H and bending moment per unit length M at the line of contact of dome and ring beam.

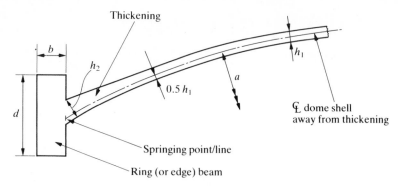

Figure 1.1

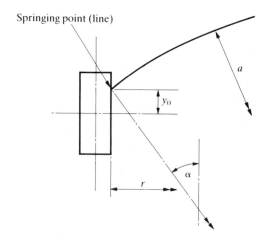

Figure 1.2

1.3 Design Formulae for Non-prestressed Domes

1.3.1 Analytical Process

This section sets out considerations that should be made prior to the design of a non-prestressed dome and details the analyses that should be made at various stages in the design.

1   The shell of a dome is considered as supporting its loading by inplanar, or membrane, forces or stresses. The edge reactions necessary for

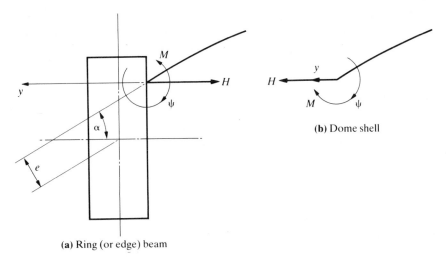

(b) Dome shell

(a) Ring (or edge) beam

Figure 1.3

equilibrium can be considered in terms of their vertical and horizontal components. The former are carried determinately by the ring beams which are supported vertically (for example with the wall of a tank). The horizontal components are carried by the ring beam which usually stretches, allowing movement of the shell edge by a different amount than that associated with the membrane forces in the shell. The ring beam will also rotate. The membrane forces are determined in Section 1.3.2 for a loading of q per unit of curved surface area. In practice the dead and live loads are usually considered in this way to simplify and speed design. The self-weight of the dome shell, although thickened locally near the ring beam, is approximated in this way. Furthermore, consideration must be given to both snow and wind loads. For the UK the snow load is specified as uniform per plan area. Taking this as per curved surface area is conservative, and no less wrong,[1.12] as snow tends to be blown away from the crown and piles higher towards the ring beam; i.e. the distribution of snow load specified by the code is only a loose approximation. Wind loads are generally suction and do not overcome the self-weight so that they are usually ignored in practical design.

2 The ring beam moves horizontally and rotates. These movements at the line of contact with the shell edge are generally different to those calculated for the shell edge by the membrane theory.

3 Unknown equal and opposite moments and horizontal forces, see Fig. 1.3, have to be considered between the shell and edge beam at their line of contact.

4 To determine these unknown moments and forces the rotations and horizontal deflections have to be equated, and to do this it is necessary to determine the rotations and deflections in Section 1.3.3.

1.3.2 Membrane Stress Resultants for Shell

These are internal membrane forces per unit length N_ϕ and N_θ due to loading q per unit area of curved surface (as shown in Fig. 1.4).

$$N_\phi = -\frac{aq}{1 + \cos\phi} = (N1) \tag{1.1}$$

$$N_\theta = aq\left(\frac{1}{1 + \cos\phi} - \cos\phi\right) = (N2) \tag{1.2}$$

where a is the radius of the shell at the point considered. These formulae are the same as given in Refs 1.1–1.3, 1.6–1.8. At the edge $\phi = \alpha$.

1.3.3 Displacements and Rotations due to M and H

These are required for the shell edges and for the ring beam at its line of contact with the shell.

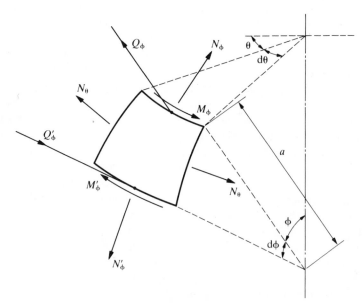

Figure 1.4

Dome edges   The directions used for H, y, M and ψ are as shown in Fig. 1.3(b).

$$y = \frac{(C1)}{E_c} M + \frac{(C2)}{E_c} H = \frac{(Y1)}{E_c} \tag{1.3}$$

where

$$(C1) = \frac{2\lambda^2}{h} \sin \alpha \tag{1.4}$$

$$(C2) = \frac{2a\lambda}{h} \sin^2 \alpha \tag{1.5}$$

$$\lambda = 4\sqrt{3(1 - v^2)\left(\frac{a}{h}\right)^2} = (L) \tag{1.6}$$

$$v = \text{Poisson's ratio} = (V) \tag{1.7}$$

Thus

$$(Y1) = (C1)M + (C2)H \tag{1.8}$$

$$\psi = \frac{(C3)}{E_c} M + \frac{(C4)}{E_c} H = \frac{(F1)}{E_c} \tag{1.9}$$

where

$$(C3) = \frac{4\lambda^3}{ha} \tag{1.10}$$

$$(C4) = \frac{2\lambda^2}{h} \sin \alpha \tag{1.11}$$

Thus

$$(F1) = (C3)M + (C4)H \tag{1.12}$$

These formulae are the same as given in Refs 1.1, 1.3 and 1.7, sometimes with notation sign changes.

Ring beam   At springing line.

$$y = -\frac{r^2 H}{E_c bd} + \frac{12r^2 y_0}{bd^3 E_c}(-Hy_0 + M) \tag{1.13}$$

$$= (C5)\frac{H}{E_c} + (C6)\frac{M}{E_c} = (Y2)/E_c \tag{1.14}$$

$$\therefore \quad (Y2) = (C5)H + (C6)M \tag{1.15}$$

where

$$(C5) = -\frac{r^2}{bd}\left(1 + \frac{12y_0^2}{d^2}\right) \tag{1.16}$$

$$(C6) = \frac{12r^2 y_0}{bd^3} \tag{1.17}$$

$$\psi = \frac{12r^2}{bd^3 E_c}(Hy_0 - M) \tag{1.18}$$

$$= \frac{(C7)}{E_c}(Hy_0 - M) = \frac{(F2)}{E_c} \tag{1.19}$$

$$\therefore \quad (F2) = (C6)H - (C7)M \tag{1.20}$$

where

$$(C7) = (C6)/y_0 \tag{1.21}$$

1.3.4 Displacements and Rotations due to Membrane Forces

These are required for the shell edge and the for the ring beam at its line of contact with the shell.

Dome Edge

$$y = \frac{a\sin\alpha}{E_c h}(N_\theta - vN_\phi) = \frac{(Y3)}{E_c} \tag{1.22}$$

where

$$(Y3) = \frac{a^2 q \sin\alpha}{h}\left(\frac{1+v}{1+\cos\alpha} - \cos\alpha\right) \tag{1.23}$$

$$\psi = \frac{aq}{E_c h}(2+v)\sin\alpha = \frac{(F3)}{E_c} \tag{1.24}$$

These formulae agree with those given in Refs 1.2 and 1.7.

Ring Beam   At springing line.

$$y = \frac{r^2}{E_c bd}\left(\frac{aq\cos\alpha}{1+\cos\alpha}\right) + \frac{12r^2 y_0 eaq}{bd^3 E_c(1+\cos\alpha)} \tag{1.25}$$

or

$$\frac{(Y4)}{E_c} = \frac{(C8)}{E_c} + \frac{(C9)}{E_c} \tag{1.26}$$

$$\therefore \quad (Y4) = (C8) + (C9) \tag{1.27}$$

where

$$(C8) = \frac{r^2 aq \cos \alpha}{bd(1 + \cos \alpha)} \tag{1.28}$$

$$(C9) = \frac{12r^2 y_0 eaq}{bd^3(1 + \cos \alpha)} \tag{1.29}$$

$$\psi = -\frac{12r^2 aqe}{bd^3 E_c(1 + \cos \alpha)} = \frac{(F4)}{E_c} \tag{1.30}$$

and e is as shown in Fig. 1.5. From this figure

$$e = \left(y_0 - \frac{b}{2} \tan \alpha \right) \cos \alpha$$

$$= y_0 \cos \alpha - \frac{b}{2} \sin \alpha \tag{1.31}$$

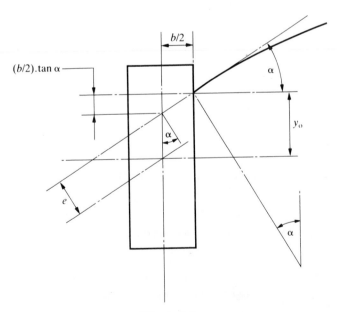

Figure 1.5

1.3.5 Compatibility Equations

To obtain the unknowns H and M the deflections and rotations of the shell edge and the ring beam, where connected to the shell, are equated. Thus

$$(Y3) + (Y1) = (Y4) + (Y2) \qquad (1.32)$$

$$(F3) + (F1) = (F4) + (F2) \qquad (1.33)$$

These give

$$M[(C1) - (C6)] + H[(C2) - (C5)] = -(Y3) + (Y4) \qquad (1.34)$$

$$M[(C3) - (C7)] + H[(C4) - (C6)] = (F4) - (F3) \qquad (1.35)$$

and these two equations can be solved for M and H.

1.3.6 Forces and Moments in Shell

For design purposes, internal membrane forces in the shell part of the dome are obtained from Eqs (1.1) and (1.2) and these are added to the internal forces and moments, due to the values of H and M determined as in Section 1.3.5, obtained from the following formulae which generally agree with Refs 1.2 and 1.7)

$$N_\phi = -\sqrt{2}\sin\alpha \exp(-\lambda\phi)\sin(\lambda\phi - \pi/4)\,H/\tan(\alpha - \phi) \qquad (1.36)$$

$$N_\theta = -2\lambda\sin\alpha \exp(-\lambda\phi)\sin(\lambda\phi - \pi/2)\,H \qquad (1.37)$$

$$M_\phi = a\sin(\alpha)\exp(-\lambda\phi)\sin(\lambda\phi)\,H/\lambda \qquad (1.38)$$

$$Q_\phi = \sqrt{2}\exp(-\lambda\phi)\sin\alpha\sin(\lambda\phi - \pi/4)\,H \qquad (1.39)$$

$$N_\phi = -2\lambda\exp(-\lambda\phi)\sin(\lambda\phi)\cot(\alpha - \phi)\,M/a \qquad (1.40)$$

$$N_\theta = -2\sqrt{2}\lambda^2\exp(-\lambda\phi)\sin(\lambda\phi - \pi/4)\,M/a \qquad (1.41)$$

$$M_\phi = \sqrt{2}\exp(-\lambda\phi)\sin(\lambda\phi + \pi/4)\,M \qquad (1.42)$$

$$Q_\phi = 2\lambda\exp(-\lambda\phi)\sin(\lambda\phi)\,M/a \qquad (1.43)$$

1.3.7 Forces and Moments in Ring Beam

The ring beam has to resist the internal forces and moments, due to the values of H and M determined as in Section 1.3.5, and the internal membrane force from Eq. (1.1) at the edge of the shell. However, Appendix 5 gives values of N_ϕ, Q_ϕ and M_ϕ which have taken into account membrane forces

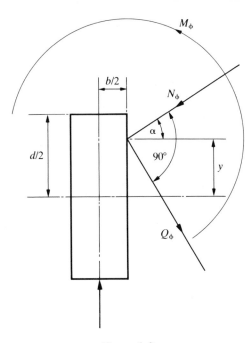

Figure 1.6

and the values of H and M at the edge of the shell. These, therefore, have to be carried by the ring beam as shown in Fig. 1.6. The forces on Fig. 1.6 can be resolved into horizontal H_r and vertical V_r forces at the centroid of the ring beam and a couple M_r about the centroid, as shown in Fig. 1.7, where

$$V_r = N_\phi \sin \alpha + Q_\phi \cos \alpha \tag{1.44}$$

$$H_r = N_\phi \cos \alpha - Q_\phi \sin \alpha \tag{1.45}$$

$$M_r = M_\phi - V_r \frac{b}{2} + H_r y \tag{1.46}$$

From Billington,[1.7] H_r causes an axial tension in the ring beam of T_r where

$$T_r = H_r r \tag{1.47}$$

and M_r causes a bending moment about the horizontal axis of the ring beam of M_{rx} where

$$M_{rx} = M_r r \tag{1.48}$$

When M_{rx} is positive it causes the top of the ring beam to be in tension and the bottom of the ring beam to be in compression.

9

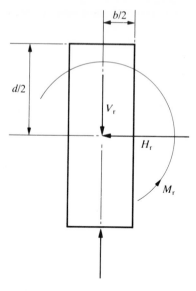

Figure 1.7

The tensile stress caused by T_r is $T_r/(bd)$ for unit loading per unit area of shell.

In Eqs (1.44) and (1.45) the values of N_ϕ and Q_ϕ are obtained for unit loading per unit area of shell.

The stresses due to the moment M_{rx} are superimposed upon the uniform tensile stress due to T_r. The tensile stress in the top fibres of the beam due to M_{rx} equals $6\,M_{rx}/(b\,d^2)$, and the magnitude of the compressive stress in the bottom fibres of the beam due to M_{rx} is the same. Therefore, the top fibre stress f_t, shown as ft in Appendix 4, is given by

$$f_t = \frac{T_r}{b\,d} - \frac{6\,M_{rx}}{b\,d^2} \tag{1.49}$$

where a positive value for f_t denotes tensile stress and the bottom fibre stress f_b, shown as fb in Appendix 4, is given by

$$f_b = \frac{T_r}{b\,d} + \frac{6\,M_{rx}}{b\,d^2} \tag{1.50}$$

where a positive value for f_b denotes tensile stress.

1.4 Checks on Non-prestressed Design Tables (Appendix 5)

1.4.1 At the Shell Edge

From Appendix 5 the forces N_ϕ and Q_ϕ were resolved horizontally and vertically, and the M_ϕ moment was also used to assess the values of f_t and f_b on the ring beam.

To check these values, the values of the indeterminate force H and moment M (obtained from the solution of simultaneous equations) and the horizontal and vertical components of the shell edge membrane force N_ϕ were used to assess f_t and f_b on the ring beam.

These agreed.

1.4.2 Stresses in Ring Beam

The stresses f_t and f_b, from Appendix 5, were used to assess the stress in the ring beam at the springing line of the shell part of the dome.

This was compared with the stress at the edge of the shell using N_θ from Appendix 5 and dividing it by h.

These agreed.

1.5 Design Formulae for Prestressed Domes

1.5.1 Analytical Process

A prestressing force whose resultant is a horizontal force whose line of action passes through the centroid of the cross-section of the ring beam and compresses the latter, is considered.

Unlike the analysis of a non-prestressed dome (Section 1.3.7) the prestressing force also causes a horizontal displacement of the ring beam.

1.5.2 Displacement and Rotations of Ring Beam

At the springing line displacement and rotations of the ring beam occur due to prestressing. For a prestressing force of 1 kN

$$y = - \frac{r^2}{E_c b(r + b)d} = \frac{(Y5)}{E_c} \tag{1.51}$$

$$\therefore \quad (Y5) = - \frac{r^2}{b(r + b)d} \tag{1.52}$$

$$\psi = 0 \tag{1.53}$$

11

1.5.3 Compatibility Equations

Under prestressed conditions the compatibility equations of Section 1.3.5 for the prestressing force only, and not the membrane forces, become

$$(Y1) = (Y2) + (Y5) \qquad (1.54)$$

$$(F1) = (F2) \qquad (1.55)$$

These give

$$M[(C1) - (C6)] + H[(C2) - (C5)] = (Y5) \qquad (1.56)$$

$$M[(C3) - (C7)] + H[(C4) - (C6)] = 0 \qquad (1.57)$$

and these two equations can be solved for M and H.

1.5.4 Resultants

For the prestressing force only the values of M and H from Section 1.5.3 are used, as in Sections 1.3.6 and 1.3.7, to obtain forces, stresses and moments in the shell and its ring beam. These are given in Appendix 6, and are then superimposed with the results given in Appendix 5; i.e. for the same dome but not prestressed. For example, for a non-prestressed dome use the results of say Table 8A, p 80, multiplied by the loading required (as the table is for a load of $1 \, \mathrm{kN \, m^{-1}}$ of curved surface area); then for a prestressed dome of the same dimensions and loading, superimpose upon these results the results of Table 8AP, p 200, multiplied by the prestressing force (as the table is for a prestressing force of 1 kN).

1.6 Checks on Prestressed Design Tables (Appendix 6)

The stresses f_t and f_b from several of the tables given in Appendix 6 were used to assess the stresses in the ring beam at the springing line of the shell part of the dome.

This was compared with the stress at the edge of the shell using N_θ from Appendix 6 and dividing it by h.

These agreed.

References

1.1 Timoshenko S and Woinowsky-Krieger S 1959 *Theory of plates and shells.* New York, McGraw-Hill

1.2   Chatterjee B K 1971 *Theory and design of concrete shells.* London, Edward Arnold

1.3   Ramaswamy G S 1968 *Design and construction of concrete shell roofs.* New York, McGraw-Hill

1.4   Flügge W 1973 *Stresses in shells.* New York, Springer-Verlag

1.5   Kraus H 1967 *Thin elastic shells.* New York, Wiley

1.6   Schnobrich W C 1974 Thin shell structures. In M Fintel (ed) *Handbook of concrete engineering.* New York, Van Nostrand Reinhold

1.7   Billington D P 1965 *Thin shell concrete structures.* New York, McGraw-Hill

1.8   Wilby C B 1991 *Concrete materials and structures.* Cambridge–New York, Cambridge University Press

1.9   Wilby C B 1980 *Design graphs for concrete shell roots.* London, Applied Science (Elsevier) Publishers

1.10   Wilby C B and Khwaja I 1977 *Concrete shell roofs.* London, Applied Science (Elsevier) Publishers

1.11   Wilby C B and Naqvi M M 1973 *Reinforced concrete conoidal shell roofs.* London, Cement and Concrete Association; now, Andover, Chapman & Hall Ltd

1.12   Parkinson J 1982 Roof shape is key to snow code. *New Civil Engineer*: 21 January

2 Calculation of Design Tables

2.1 Introduction

This chapter assesses: the recognized values of loading (both dead and live), requisite dome shell thicknesses, concrete mix specifications for the reinforcement cover, and the physical properties and mechanical behaviour of the reinforcement that are given in authoritative publications. Limiting conditions are established for the design tables of Appendices 5 and 6, and for the examples given in Chapter 3.

Sections 2.7 and 2.8 indicate how the tables presented in Appendices 5 and 6 can be used for the design of non-prestressed and prestressed domes respectively.

2.2 Loading

It has been established in the Preface and Section 2.6 that elastic analysis is desirable in the design of dome roofs. This suggestion is supported by Ref. 2.1 and by ACI 318–89. The latter also allows permissible stress/modular ratio elasticity design of sections, as does BS 5337:1976 for water-retaining structures. One of the main factors to influence the design is the loading.

The finishes of the dome shell contribute to the loading, and in the British Isles most shells consist of three layers of built-up roofing felt (the top one mineral finished) on top of either 25 mm thick cork board or 50 mm thick vermiculite lightweight concrete. The soffit (i.e. the underside) of the shell can be painted, or plastered then painted. Wind and snow loadings also contribute to the overall value, and further information on these is given in Appendices 1 and 2 respectively.

Suggested loadings for the UK are:

superimposed
snow	0.750
wind	–
roofing felt	0.072
50 mm vermiculite	0.290
plaster and paint	0.093
TOTAL LOADING	1.205 kN m^{-2}

Suggested loadings for much of the USA are:

superimposed

snow	20 psf = 0.958
wind	–
roofing felt	0.072
25 mm cork board	0.060
plaster and paint	0.093
	1.183 kN m^{-2}
TOTAL LOADING	say 1.205 kN m^{-2}

A useful loading to use in the examples in this book is, therefore, 1.205 kN m^{-2} (i.e. 25.17 psf). The tables in Appendices 5 and 6 can of course be used for any value of loading.

2.3 Buckling

Failure of a dome construction may be initiated by buckling of the shell. Realistic tests using reinforced concrete domes are difficult and expensive to make; however, it is generally accepted amongst investigators that a popular formula for load buckling is given by

$$\text{buckling loading} = CE_c(h/a)^2 \qquad (2.1)$$

where E_c = the modulus of elasticity (Young's modulus) of the concrete and C is given different values by various investigators.

To be on the safe side, i.e. to allow for an apparent reduction in E_c due to long-range creep (plastic flow), it becomes a problem deciding what value to take for E_c. The shell concrete is normally only required to resist low stresses, thus a very strong concrete with a high E_c is not used. A further point is that a considerable proportion of the loading is dead, as opposed to live loading, and is, therefore, always present helping creep. Again it has been reported (pers com) that certain cylindrical shell roofs eventually, after about twelve years, increased their initial deflection by about three times. These shells probably never experienced their live loading, and to investigate this defect a great number of tests in both laboratory and on site have been conducted. Measured deflections on site, of the kind of reinforced concrete of which the dome would be constructed, have indicated a Young's modulus for the concrete of about one-tenth that of steel; i.e. the modular ratio would be 10. The British code of practice, CP114, dealing with elastic design, recommends a modular ratio of 15. This allows for long-range creep, and a test in the laboratory on a similar concrete at the age of about 28 days would probably give a modular ratio of 10. Using British codes, BS 8110 gives the

15

modulus of elasticity of reinforcement as $200\,\mathrm{kN\,mm^{-2}}$. Using CP110

$$E_c = 200/15 = 13.33\,\mathrm{kN\,mm^{-2}} = 13\,333\,000\,\mathrm{kN\,m^{-2}}$$

$$= 1\,933\,700\,\mathrm{psi}$$

ACI 318–89 requires that the specified compressive strength of concrete (obtained by testing USA cylinders at 28 days) for shells should not be less than 3000 psi (equivalent to a British cube strength of about $3000/0.84 = 3571$ psi) and that for normal weight concrete

$$E_c = 57\,000\sqrt{3000} = 3\,122\,000\,\mathrm{psi} = 21\,526\,000\,\mathrm{kN\,m^{-2}}$$

Seemingly not as much creep is allowed for as in CP110.

So, for safety (and to be backed by a code, CP110), E_c is taken as $13\,333\,000\,\mathrm{kN\,m^{-2}}$. Should the reader consider it necessary to use a different value of E_c, due to different code regulations, then the quantity qb shown in Appendices 5 and 6 (the proposed safe total loading, q, on the dome to guard against buckling) is adjusted to his preferred new value of E_c.

C is given different values by various investigators. Ramaswami,[2.2] quotes Schmidt and Csonka who recommend values of C of 0.15 and 0.06 respectively. Professor Haas of Delft also refers to the latter. Schnobrich suggests a value of 0.175. Ramaswami also quotes the Indian standard 2210 as giving $C = 0.10$. Appendices 5 and 6 use the conservative value of $C = 0.06$.

The British codes have used a load factor of 1.8 for reinforced concrete design and this allows for imperfections and cracking. These latter may have a worse effect in the case of domes and a greater load factor is desired against buckling to avoid this effect precipitating failure, as it cannot be predicted as accurately as other elements of the design. Appendices 5 and 6, therefore, use a load factor of 3. Should the reader consider it necessary to use a different value to this from his studies or national code then the quantity qb, the proposed safe total loading, is adjusted to his preferred value.

Therefore, from Eq. 2.1

$$qb = 0.06 \times 13\,333\,000 \times (h/a)^2/3$$

$$= 266\,660\,(h/a)^2\,\mathrm{kN\,m^{-2}} \tag{2.2}$$

ACI 318R–89 refers to Ref. 2.3. This gives the same formula as Eq. 2.1 but gives C as 0.25 and E_c as $58\,000\sqrt{3000} = 3\,176\,800$ psi, using 3000 psi as the cylinder concrete strength and 58 000 in lieu of the 57 000 quoted previously. But then as the E_c is an initial elastic modulus not allowing for creep, and as shrinkage has not been allowed for, it cites load factors of 4–6 and considers that it should be at least 4 for domes covering storage areas and 'considerably more for public buildings'. Assuming a value of 6 for the load factor in Eq. 2.1, ACI 318R–89 gives

$$qb = 0.25 \times 3\,176\,800 \times (h/a)^2/6 = 132\,370\,(h/a)^2 \text{ psi}$$

$$= 912\,700\,(h/a)^2 \text{ kN m}^{-2} \tag{2.3}$$

Zarghamee and Heger,[2.4] consider that imperfections of construction reduce the buckling load by 50 per cent, such that the value of C becomes 0.33. They consider that 'the effects of flexural cracking on the buckling strength of the dome to be insignificant'.

The investigators mentioned previously in this section have generally considered the IASS recommendations.[2.5]

Publications generally do not seem to distinguish between domes with and without prestressed ring beams with regard to resistance to buckling.

Applying Eq. 2.2 to many domes constructed in the past, indicate their inadequacy, sometimes considerable, in buckling. These have not seemingly failed but may never have experienced their full design loading. In addition, Eq. 2.2 is very much more conservative than Eq. 2.3 which enjoys the authority of the ACI. Therefore, for Appendices 5 and 6, the load factor used in Eq. 2.2 is reduced from 3 to 2.2 (which is still greater than the 1.8 mentioned as having been used by British codes) to give a value of $qb = 399\,990 \times (h/a)^2 \text{ kN m}^{-2}$. Of course designers wishing to take a more conservative view should make their own assessment of buckling when using Appendices 5 and 6.

When the shell of the dome is thickened towards its thickening, the weight of the thickening is not over the crown but in a position much less likely to cause buckling and is considered in Section 2.4 below.

2.4 Thickness of Shell Part of Dome

Deciding upon the shell thickness of the dome presents difficulties just as it does with cylindrical, hypar and other shapes of structural concrete shells.

Everyone seems to agree that the thin shell which can be used needs thickening towards, and when it reaches, the ring beam. This is because there is a sudden change of structural system assisting vulnerability to cracking, due to stress concentrations of the kind that can be observed in photoelastic tests on structures generally and which are ignored in analyses, and practical shrinkage and temperature movements. In addition the thickening helps reduce the reinforcement required to resist the bending moment M_ϕ (Mf in Appendix 4) and helps resist the shear forces due to the rate of change of this bending moment, i.e. Q_ϕ (Qf in Appendix 4). Again, there is difficulty deciding upon the thickness and length from the springing point for this thickening.

ACI 318–89 requires that thickness shall be proportional to the required strength and serviceability. Generally, in practice very thin shells can adequately resist the forces and moments, from an analysis, and buckling.

ACI 344R – 70 (reapproved 1981) recommends that the dome shell thickness should not be less than 3 in (c 80 mm) for concrete and 2.5 in (c 65 mm) for shotcrete domes.

Faber and Mead,[2.6] report that the shell thickness has been as little as 1/650 of the span and that a ratio of 1/450 has been common. This latter meant that a 65 mm (2.5 in) thick shell can span 29.250 m (c 96 ft) and an 80 mm (c 3 in) thick shell 36 m (c 118 ft). In Appendices 5 and 6 the ratio 1/450 is not exceeded.

With regard to domes with prestressed ring beams Dobell[2.7] offers considerable guidance, and surprisingly mentions that 'many have been built with only 2 in (c 51 mm) thickness for spans up to 150 ft (c 45.7 m)' but adds that such thin shells 'have a tendency to crack ... and it is now the usual practice to use a thickness varying from 2 in (c 51 mm) to 6 in (c 152 mm) for diameters of 50 ft (c 15.2 m) to 250 ft (c 76.2 m)'. In a table he cites 2 in (c 51 mm), 2.5 in (c 64 mm), 4 in (c 102 mm) and 4.5 in (c 114 mm) thick shells for spans up to 101 ft (c 30.8 m), 120 ft (c 36.6 m), 151 ft (c 46.0 m) and 190 ft (c 57.9 m) respectively.

Zarghamee and Heger[2.4] have surveyed existing prestressed concrete domes in the USA and their frequency distribution shows thicknesses ranging between 2 in (c 76 mm) and 5 in (c 127 mm) with a very significant majority having a thickness of 3 in (c 76 mm).

Generally the thinner the sheli the more economic because of the self-weight reduction, even though some extra reinforcement may be required.

The references cited in this book do not seem to distinguish between the shell thickness of a non-prestressed dome and that of a prestressed dome. The shell itself is not prestressed in the latter case, but it might be argued that the moments and forces in it are less. These are, however, influenced by the geometry of the ring beam which may be such that the last-mentioned moments and forces are increased for prestressed domes due to the use of smaller ring beams.

Generally, for the practical shell thicknesses used the stresses in the shell due to moments and forces are low, whether or not the dome is prestressed, and in either case the shell will be designed to be satisfactory against buckling.

After about 1945 prestressed concrete construction and cylindrical shell roofs 2.5 in (64 mm) thick started to be used in the UK. Prior to this domes constructed in the UK, in buildings and over tanks, were generally much thicker.

The 1976 PCA leaflet 'Design of circular domes' makes no distinction between prestressed and non-prestressed domes and says '3.5 in (89 mm) is about as thin as any shell can be made'! Most references studied seem to use thinner shells for prestressed than for non-prestressed domes. However,

Table 2.1

Span (plan diameter)	Shell thickness
12–18 m (c 40–60 ft)	75 mm (c 3 in)
>18–24 m (c 80 ft)	90 mm (c 3.5 in)
>24–30 m (c 100 ft)	100 mm (c 4 in)
>30–36.1 m (c 120 ft)	130 mm (c 5 in)

Table 2.2

Span (plan diameter)	Springing thickness
12–18 m (c 40–60 ft)	180 mm (c 7 in)
>18–24 m (c 80 ft)	200 mm (c 8 in)
>24–30 m (c 100 ft)	250 mm (c 10 in)
>30–36.1 m (c 120 ft)	290 mm (c 11.5 in)

several of these domes have inadequate resistance to buckling according to Section 2.3.

For the tables given in Appendices 5 and 6 it would seem sensible to take notice of the ACI 344R–70 (reapproved 1981) recommendation that the dome shell thickness should not be less than 3 in and use 75 mm (2.953 in) as a minimum thickness. For larger domes the thickness was limited by the buckling considerations of Section 2.3, whether or not the dome has a prestressed ring beam. Thus the shell thicknesses given in Table 2.1 have been used for domes, whether or not prestressed.

As described previously many references agree that the thin shells which can be used need thickening towards, and when they reach, the ring beams. None appear to recommend dimensions. Reference 2.8 recommends dimensions for cylindrical shell roofs as ranging between 130 (c 5 in) and 230 mm (c 9 in) and has the span divided by 120 which is also the width divided by 60. These thickenings seem rather greater than used for various domes. In an example by Billington[2.9] the thickness to the ring beam is increased 2.4 times. It was, therefore, decided to use the values shown in Table 2.2 for the thicknesses of the domes in Appendices 5 and 6.

For the tables given in Appendices 5 and 6 the varying shell thickness of a dome is dealt with analytically as recommended by Billington[2.9] using the paper by Hanna.[2.10]

The method first decides that the distance, S, on the curved surface of the shell from which the thickened region extends should be given as

$$S = 2\sqrt{ah_a} \qquad (2.4)$$

where a is the radius of the shell and h_a is the average thickness of the

19

thickened region, i.e. $(h_1 + h_2)/2$, where h_1 is the thin part of the shell over the crown, etc. and h_2 is the thickest part of the shell at its junction with the ring beam.

Then the method considers, for the purpose of the analysis, each shell to have an equivalent constant thickness, h, given by

$$h = 0.75h_2 + 0.25h_1 \qquad (2.5)$$

The analyses are made for Appendices 5 and 6 for a total loading of curved surface area of $1\,\mathrm{kN\,m^{-2}}$. The values of internal forces and moments given in the tables have, therefore, to be multiplied by the loading, in $\mathrm{kN\,m^{-2}}$ of curved surface area, that the designer considers appropriate.

The thickening poses a problem in how to account for its self-weight.

Chatterjee,[2.11] in the only example he gives, which is for a 3 in (76 mm) shell thickened to 6 in (152 mm), allows 14 psf ($0.67\,\mathrm{kN\,m^{-2}}$) for waterproofing and thickening, considered as spread uniformly over the curved surface area of the shell, and adds this to the weight of the 3 in (76 mm) thickness which he calls 36 psf ($1.72\,\mathrm{kN\,m^{-2}}$). It is assumed, in this calculation, that the waterproofing consists, say, of three layers of built-up roofing felt weighing 1 psf ($0.05\,\mathrm{kN\,m^{-2}}$) and that no allowance has been made for contributions to the loading from vermiculite, cork board or any other insulation. Thus, according to Chatterjee

loading due to self-weight of shell + thickening

$$= 36 + 13$$

$$= 49\ \mathrm{psf}\ (2.35\,\mathrm{kN\,m^{-2}})$$

However, a more accurate calculation of the weight taking into account the thickening and spreading it over the curved surface gives 6.62 psf, i.e. $0.32\,\mathrm{kN\,m^{-2}}$. Thus for Chatterjee

true loading due to self-weight of shell + thickening

$$= 36 + 6.62$$

$$= 42.68\ \mathrm{psf}\ (2.04\,\mathrm{kN\,m^{-2}})$$

Furthermore, Chatterjee takes $h = 3$ in (76 mm), uniformly, for his analysis, i.e. thus not allowing for the effect of the thickening. Of course, many cylindrical shells have been designed making the same assumption, except that the weight of the thickening is assumed to be carried directly at the valley edge support and its weight is not spread out over the shell (refer to publications by C B Wilby, e.g. Ref. 2.12).

Billington,[2.9] in the only example he gives, which is for a 2.5 in (64 mm) shell thickened to 6 in (152 mm) at the springing line of the dome, analyses this as though of a uniform thickness of 4 in (102 mm) and for the design, the self-weight of 50 psf ($2.39\,\text{kN m}^{-2}$) he uses for the shell, corresponds to this thickness. He initially calculates h from Eq. 2.5 as 5.12 in (130 mm) then decides, but gives no reason for this, to take 4 in (102 mm) for his analysis. However, a more accurate calculation of the weight of the thickening and spreading it over the curved surface would give 9.92 psf ($0.48\,\text{kN m}^{-2}$). Thus, for Billington, on this basis

$$\text{true total loading} = 31.25 + 9.92$$

$$= 41.17\,\text{psf}\,(1.97\,\text{kN m}^{-2})$$

[NB Chatterjee uses a weight density of $144\,\text{lb/ft}^3$ ($22.6\,\text{kN m}^{-3}$) and Billington uses $150\,\text{lb/ft}^3$ ($23.6\,\text{kN m}^{-3}$).]

There is, therefore, a considerable difference between the design methods of Chatterjee and Billington. In both cases the weight of the thickening is unduly overestimated for want of an accurate calculation of its weight.

Billington argues that if a greater thickness for h is used in the analysis, the shell part of the dome becomes stiffer and takes a larger proportion of, for example, the horizontal thrust provided by the ring beam resulting in it taking more ring tension, N_θ. The ring beam then being less stiff by comparison would need to resist less tension.

The M_ϕ moments at and towards the springing line are thus of prime relevance and the reinforcement required by these moments in this zone can be prohibitive to the design of a dome, particularly as the effective depth of the reinforcement is so small beyond and just before the end of the thickening.

Therefore, it seems desirable to consider the effects of the thickening (which is desired to reduce cracking and reinforcement to resist M_ϕ) in the analysis by using Eq. 2.5, and to not excessively and undesirably overestimate the weight of the thickening, since the thickening is a minor but not insignificant proportion of the total loading. For example, assuming a live load plus at least waterproofing of 16 psf (c $0.8\,\text{kN m}^{-2}$), for Chatterjee

$$\text{thickening, as a proportion, of maximum total load} = \frac{100 \times 6.62}{42.68 + 16}$$

$$= 11.3\%$$

For Billington

$$\text{thickening as a proportion, of minimum total load} = \frac{100 \times 9.92}{41.17 + 16}$$

$$= 17.4\%$$

So to consider the weight of the perimeter thickenings as spread uniformly over the whole dome shell, undesirably and unfairly overestimates the maximum values of the M_ϕ moments, and this is made worse if the weight of the thickenings is overestimated as cited previously.

Thus in Appendices 5 and 6 the weight of the thickening has been calculated accurately from spherical coordinate geometry, and has, after consideration, been spread uniformly over the curved surface of the shell. This is a compromise between the previously mentioned practices of Chatterjee and Billington. It is, however, prudent to spread the load in this way, for example with regard to M_ϕ moments in the shell.

The weight density of reinforced concrete has been taken as $23.6 \, \text{kN} \, \text{m}^{-3}$ (150.3 lb/cu ft) for Appendices 5 and 6.

The volume of the thickening can be shown to be equal to

$$(h_2 - h_1)S\pi a \, \sin(\alpha - S/3/a) \tag{2.6}$$

The surface area of the shell can be shown to be equal to

$$2\pi a^2(1 - \cos \alpha) \tag{2.7}$$

For example, if the span $= 30 \, \text{m}$, $h_1 = 100 \, \text{mm}$, $h_2 = 250 \, \text{mm}$, and $\alpha = 28$ degrees, then $h_a = 175 \, \text{mm}$, $S = 4.729 \, \text{m}$, the volume of the thickening $= 30.29 \, \text{m}$ and the surface area of the shell $= 750.9 \, \text{m}^2$. Taking this volume and dividing it by the surface area and adding h_1 gives an equivalent shell thickness for weight purposes of $140 \, \text{mm}$ and using a weight density for the reinforced concrete of $23.6 \, \text{kN} \, \text{m}^{-3}$ a value of the self-weight of the shell of $3.312 \, \text{kN} \, \text{m}^{-2}$ (69.2 psf) is obtained. In Appendix 4 this is denoted as qD. Adding this to the weight of the live load and finishes, per m^2, gives a value of total loading, per m^2, for use for multiplying by the values of forces and moments that are given in Appendices 5 and 6 for a unit loading of $1 \, \text{kN} \, \text{m}^{-2}$.

Now in Section 2.3 a safe buckling load was suggested appertaining to a shell of uniform thickness. This latter is shown as qb in Appendix 4. For buckling considerations it would seem unfair to spread the weight of the thickening at the edges over the whole curved surface area, because the weight at the crown is much more influential on buckling than weight at the perimeter.

For this reason only 30 per cent of the weight of the thickening was spread over the curved surface area, giving a loading for the self-weight of the shell and its thickening shown in Appendix 4 as qd. This is added to live loading, and loading due to waterproofing, soffit finishes and insulation, etc. to check that the total is less than the safe buckling load. In this case, in the example, taking 30 per cent of the volume and dividing it by the surface area and adding h gives an equivalent shell thickness for weight purposes of $112.1 \, \text{mm}$, and using a weight density of $23.6 \, \text{kN} \, \text{m}^{-3}$ a value of $2.646 \, \text{kN} \, \text{m}^{-2}$ (55.26 psi) is obtained for qd.

If one wished to take all the weight of the thickening as spread over the shell, then one would use qD on the tables in lieu of qd. As the thickening is not over the crown, some designers might just take h and multiply it by the weight density of the shell to obtain the total loading. In the example this gives a value of the self-weight of the shell of $2.36\,\mathrm{kN\,m^{-2}}$ (49.3 psf).

2.5 Concrete and Cover to Reinforcement

When British Imperial units and CP114 were used, there was a wealth of experience in the UK concerning the use of 2.5 in (63.5 mm) thick cylindrical shells composed of a 1:2:4 concrete mix, having a cube strength not less than 3000 psi ($20.7\,\mathrm{N\,mm^{-2}}$) at an age of 28 days. Following the partial collapse (flattening out) of an N-light shell, the main UK design company changed the concrete mix specifications to 1:1.5:3. However, it was eventually found that the trouble was due to a reinforcement detailing mistake in the edge beam that had to be externally post-tensioned for rectification, and not the strength of the concrete in the shell. At that time also 0.5 in (12.7 mm) cover of concrete to the reinforcement in the shell was commonly used. This was to help keep a reasonable effective depth to the reinforcement resisting bending moments in a shell which was only 2.5 in (63.5 mm) thick. This practice has generally been satisfactory over about 40 years.

The basic requirements of cover are not to impair the bond/adhesion between the reinforcement and concrete for structural reasons, to provide protection against corrosion, and to protect for a specified time the steel from the weakening effects of a fire. The 0.5 in (12.7 mm) cover to bars with diameters less than or equal to 0.5 in (12.7 mm) proved adequate for the first two of these requirements. Fire regulations vary internationally but it was often the practice in the UK not to restrict roof design, permitting for example unclad structural steel portals, trusses, etc. to support roofs. So the 0.5 in (12.7 mm) inch cover was allowed by fire regulations.

In one example severe corrosion of the shell reinforcement occurred over about 30 years, even though the shell was well protected with roofing felt. This was proved to be due to the presence of calcium chloride (not now used) as a hardener in the concrete mix. Certain well-protected ties each had a post-tensioning cable, consisting of eight 5 mm diameter wires, running down its centre with plenty of cover. Often, six of the wires were completely corroded through and sometimes all eight.

When there are no detrimental chemicals the above practice concerning concrete quality and cover has seemingly proved satisfactory over a long period.

Subsequent years have seen a period of specifying strength of mix for concrete structures generally. Sometimes ready-mix suppliers and contractors design mixes requiring considerable compaction, because of low water/cement ratio, to gain a low cement content in the mix and thereby reduce its cost.

Table 2.3

Concrete	Nominal cover (mm)				
	25	20	15	15	15
Max free water/cement ratio	0.65	0.60	0.55	0.50	0.45
Min cement content (kg m^{-3})	275	300	325	350	400
Min characteristic compressive strength (N mm^{-2})	25	30	35	40	45

With low cement content, mixing becomes more important. Thorough mixing and compaction is not always achieved with perfection and this, and the reduced cement content, make the probability of protecting the reinforcement against corrosion sometimes inferior to when mix proportions had been specified. Seemingly for this reason CP110 and then BS 8110 specify minimum cement contents for durability against various conditions of exposure etc.

ACI 318–89 requires that shells should use concrete with a cylinder compressive strength not less than 3000 psi, which is approximately a British cube strength of 24.6 N mm^{-2}. As the same code mentions, the value of Young's modulus is increased for stronger concretes and so helps resistance to buckling.

BS 8110 and BS 5321 do not recommend the use of concretes with normal weight aggregates of characteristic strengths less than 25 N m^{-2}.

ACI 318–89 requires for shells not exposed to the weather that the minimum concrete cover to the reinforcement should be 1/2 in (c 13 mm) for a 5/8 in (c 16 mm) bar and smaller, W31 or D31 wire, and 3/4 in (c 19 mm) for a 3/4 in bar and larger.

BS 8110 for normal weight concretes states that, where the concrete surfaces are protected against weather or aggressive conditions, the nominal maximum size of aggregate should not be more than 15 mm (c 5/8 in) and if a 'systematic checking regime is established to ensure compliance with the limits' of water/cement ratio and cement content, the minimum concrete covers can be as given in Table 2.3.

Higher cement contents and rapid-hardening cements are sometimes considered to increase shrinkage and, therefore, crack widths. Of course the adverse shrinkage has to be considered relative to the rate of gain and achievement of increased tensile strength.

As shells are generally relatively quite thin it is sometimes desirable for the reinforcement cover to be as small as possible to reduce the reinforcement required to provide resistance to M_ϕ moments.

2.5.1 Poisson's Ratio

In the analyses of Appendices 5 and 6 a value of Poisson's ratio of 0.167 (i.e. 1/6) was taken. This is the same as used by Billington.[2.9]

Chatterjee[2.11] uses 0.15, Ramaswami[2.2] uses zero. Wilby[2.1] considers Poisson's ratio for concrete to be approximately between 0.20 and 0.14, dependent upon the quality of the concrete, and for steel to be approximately 0.29. Strictly speaking Poisson's ratio for the analysis should relate to reinforced concrete, the concrete being of the quality outlined in Section 2.5. An accurate value is difficult to specify, but it is often stated that its lack of accuracy has little effect on the analysis.

2.6 Reinforcement

As there is considerable satisfactory experience with elastic design and insufficient experience with limit state design Wilby[2.1] recommends using elastic analysis for determining forces and moments in domes. This is also approved by ACI 318–89.

The closest and most logical way of designing sections is, therefore, by elastic[2.1] not plastic theory. This is allowed by ACI 318–89 for shells. BS 8110 and CP110 do not use this method, but do not deliberate on shell roofs. Numerous shells have had their sections designed with satisfaction using the elastic theory of CP114.

Logically, it is most in keeping with the elastic analysis for forces and moments to use low stresses in the reinforcement. In the UK many shells have been designed using plain mild steel bars with the permissible tensile steel stresses of CP114, namely 18 000 then 20 000 psi and 140 N mm^{-2}, and have generally stood the test of time. This also applies to using high-yield steel bars with deformed (high-bond) surfaces to resist forces, and high-yield steel wire fabrics to resist bending moments in the shells; bars and wires having permissible tensile steel stresses of 27 000 then 30 000 psi and 210 N mm^{-2}. The use of high-yield steel is principally due to the fact that the major UK designer of shells, in the heyday of this kind of work, and their competitors were firms that manufactured high-yield bars and fabrics.

With high-yield steel, deflections should be greater. Also total crack widths should be increased when resisting forces. In the case of mechanically bonded bars resisting bending moments, some tests by Konji[2.13] on beams claim the well-known fact that 'at a given stress in the reinforcement the cracks will be more numerous and smaller for a mechanically bonded bar than for a plain bar'. However, the high-yield steel is used to a 50 per cent higher stress. So overall the total of crack widths should probably be greater for high-yield steel. Sometimes little reinforcement is required along lines of latitude and so using lowly stressed steel (e.g. plain bars of mild steel) assists concrete placing (i.e. provides resistance to wet concrete flowing down the sloping surface) and reduces the need to provide minimum reinforcement, see later.

For cylindrical shell roofs in the UK fabrics have been widely used to resist moments in shells; but, as fabrics are rectangular and flat they are not very easy to detail and use on a dome surface. Fabrics only have occasional mechanical bonds (where cross-wires occur), and although this is disadvantageous it does enable wires to be used at much closer centres than would be used when using reinforcement bars, and this is advantageous for bond (total bar periphery greater) and crack distribution.

In a beam one puts the reinforcement as low as possible in the tension zone, but in a shell the practice is to distribute the reinforcement according to the distribution of tensile stress. This practice was carried out by Billington[2.9] and shell designer companies in the UK[2.12–2.15] and applies to forces in the shell, but not to bending moments which are dealt with in the normal way for structural concrete.

For parts of the dome shell the analysis may indicate that tensile stresses requiring reinforcement are non-existent. On these occasions a minimum amount of reinforcement is generally provided, to resist temperature and shrinkage stresses and local stresses due, for example, to the activity of workmen. Many shells have been satisfactorily constructed over many years in the UK using the CP114 recommendation for reinforcement, of 0.15 per cent where plain bars are used, or 0.12 per cent of the gross cross-sectional area where high-yield high-bond bars or high-yield mesh are used. BS 8110 has increased these quantities to 0.24 and 0.13 per cent respectively. ACI 318–89 requires these quantities to be 0.20 and 0.18 respectively.

2.7 Use of Appendix 5 for Non-prestressed Dome Design

The design tables of Appendix 5 consider numerous domes within the ranges of the dimensions shown in Table 2.4; i.e. for these ranges of dimensions the

Table 2.4

Diameter (m)	h_1 (mm)	h_2 (mm)	b (mm)	d (mm)
36.1	130	290	300	1200
30	130	290	300	1200
29.5	100	250	300	1000
24	100	250	300	1000
23.5	90	200	280	800
18	90	200	280	800
17.6	75	180	250	700
12	75	180	250	700

plan diameter (span) has been varied in small increments so that the results can be used for any complex value of the span; for example, values such as 23.564 m and 101 ft $11\frac{5}{8}$ in.

Three types of domes, A, B and C, are dealt with by Appendices 5 (and 6) and are shown in Figs 2.1, 2.2, and 2.3 respectively. For example, Tables 6A, 8B and 4C refer to types A, B and C. Types B and C use the ring beam to provide a gutter for rainwater. Type C has a clean soffit, and thus the dome appears to be a shell without a ring beam; this design may appeal to some architects. Type C requires, if the top fibres of the ring beam are in compression, say 100 mm square struts, each reinforced with four 12 mm diameter bars and nominal stirruping, at suitable spacing for lateral instability requirements of this rectangular beam; the curvature helps in this

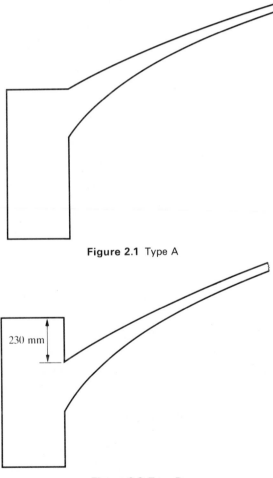

Figure 2.1 Type A

230 mm

Figure 2.2 Type B

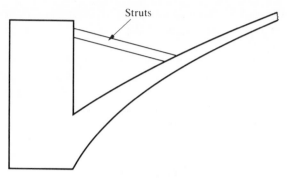

Figure 2.3 Type C

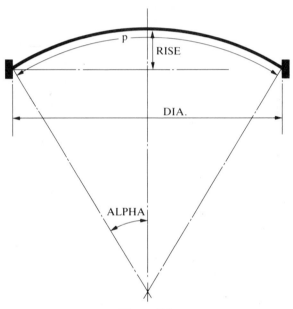

Figure 2.4

requirements of this rectangular beam; the curvature helps in this requirement. It is desirable to protect these struts with, say, roofing felt and use plenty of cover to the reinforcement. Figs. 2.4, 2.5 and 2.6 relate to Appendices 4, 5 and 6.

2.8   Use of Appendix 6 for Prestressed Dome Design

For prestressed domes the tables for non-prestressed domes are used in conjunction with tables (denoted with a P in their title) for a prestressing force of 1000 kN, whose line of action passes through the centroid of the

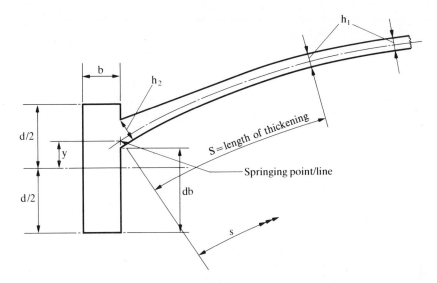

Figure 2.5

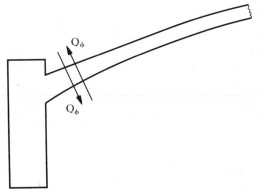

Figure 2.6

cross-section of the ring beam. For example, for a prestressed dome of certain dimensions, the table for a non-prestressed dome, say Table 6B, is used in conjunction with the corresponding table for a prestressing force of 1000 kN, namely Table 6BP. Table 6B is for a loading of $1\,\text{kN}\,\text{m}^{-2}$ and its results must be multiplied by the required loading per unit of curved surface area, in $\text{kN}\,\text{m}^{-2}$ units; whilst Table 6BP, which is for a prestressing force of 1000 kN, must have its results multiplied by the prestressing force decided upon in kN units divided by 1000 kN. The results then produced from these

29

two tables are superimposed upon one another. Reference should be made elsewhere in this book and particularly to the relevant examples.

The symbols used in Appendix 5 for a prestressing force of 1000 kN are the same as given in Appendix 4.

It will be appreciated that as in Section 2.7, types A, B and C are considered. For example, Tables 6AP, 8BP and 4CP refer to types A, B and C. Also, as in Section 2.7, type C requires the same struts as described there when the top fibres of the ring beam are in compression.

References

2.1   Wilby C B 1991 *Concrete materials and structures.* Cambridge–New York, Cambridge University Press

2.2   Ramaswami C S 1984 *Design and construction of concrete shell roofs.* Florida, Robert E. Krieger Publishing Co

2.3   ACI Committee 344, Design and construction of circular prestressed concrete structures. *ACI 344R.2R–70.* Detroit, American Concrete Institute 1970 (reaffirmed 1981)

2.4   Zarghamee M S and Heger F J 1983 Buckling of thin concrete domes. *ACI Journal,* Technical paper No 80–45, Nov–Dec

2.5   IASS Working Group No. 5 1979 *Recommendations for reinforced concrete shells and folded plates.* International Association for Shell and Spacial Structures, Madrid, Spain

2.6   Faber J and Mead F 1965 *Oscar Faber's reinforced concrete.* London, Spon

2.7   Dobell C 1951 Prestressed concrete tanks. *Proceedings first US conference on prestressed concrete.* Mass, MIT

2.8   Wilby C B 1977 *Concrete for structural engineers.* London–Boston, Mass, Newnes-Butterworth

2.9   Billington D P 1982 (and 1965) *Thin shell concrete structures.* New York, McGraw-Hill

2.10 Hanna M M 1956 Thin spherical shells under rim loading. Fifth Congress of the IABSE, Lisbon

2.11 Chatterjee B K 1988 *Theory and design of concrete shells.* London, Chapman & Hall (and 1971, London, Edward Arnold)

2.12 Wilby C B and Khwaja I 1977 *Concrete shell roofs.* Barking, Applied Science Publishers

2.13 Evans R H and Wilby C B 1963 *Concrete: plain, reinforced, prestressed and shell.* London, Edward Arnold

2.14 Wilby C B 1980 *Design graphs for concrete shell roofs.* Barking, Applied Science Publishers

2.15 Wilby C B and Naqvi M M 1973 *Reinforced concrete conoidal shell roofs: flexural theory design tables.* Andover, Chapman & Hall

3 Examples of Use of Design Tables

3.1 Non-prestressed Domes, SI Units

3.1.1 Parameters for Examples

In Section 2.6 it was established that low reinforcement stresses agree better with an elastic analysis. In the following examples elastic analysis is thus used to calculate mild steel dome reinforcement and adequacy of concrete strength, assuming: (a) a permissible stress of $140 \, \text{N} \, \text{mm}^{-2}$ (c 20 000 psi) at working loads; (b) concrete (see Section 2.5) with a cube strength not less than $25 \, \text{N} \, \text{mm}^{-1}$ (c 3000 psi USA cylinder strength) at an age of 28 days; and (c) permissible compressive stresses of $7 \, \text{N} \, \text{mm}^{-2}$ (c 1000 psi) in bending, $5.3 \, \text{N} \, \text{mm}^{-2}$ (769 psi) in direct compression and a shear stress of $0.7 \, \text{N} \, \text{mm}^{-2}$ (c 100 psi) at working loads.

Using elastic theory, since it is more in line with the elastic analysis than plastic theory, the following design formulae[3.1] are therefore used:

modular ratio equals 15
ratio of steel and concrete permissible stresses is 20
ratio of depth of neutral axis to effective depth is

$$15/(15 + 20) = 0.4286$$

percentage of reinforcement is

$$0.4286/(2 \times 20) = 1.071$$

ratio of moment (lever) arm to effective depth

$$1 - (0.4286/3) = 0.8571$$

moment of resistance/breadth/depth squared is

$$(7 \times 0.4286 \times 0.8571/2) \times 1000 = 1286 \, \text{kN} \, \text{m}^{-2}$$

Example 3.1   Design a dome of plan diameter (span) 30 m and of type A, shown in Fig. 2.1, to carry its self-weight and a loading of $1.205 \, \text{kN} \, \text{m}^{-2}$ (see Section 2.2).

Use Table 33A, where DIA. $= 30\,$m, and refer to Section 2.5 and Figs 2.4, 2.5 and 2.6. The table gives the dimensions shown on the figures for this design, and:

(a) The self-weight of the shell and 30 per cent of its thickenings (see Section 2.4) equals

$$qD = 3.4\,\text{kN}\,\text{m}^{-2}$$

Adding this to the value for total loading gives

$$\text{buckling loading} = 3.4 + 1.205$$

$$= 4.605\,\text{kN}\,\text{m}^{-2}$$

This is less than the safe buckling loading (see Section 2.3) shown on Table 33A as $qb = 6.62\,\text{kN}\,\text{m}^{-2}$, and is, therefore, satisfactory.

(b) The self-weight of the shell and of its thickenings (see Section 2.4), is $qD = 4.169\,\text{kN}\,\text{m}^{-2}$. Adding this to 1.205 gives $5.374\,\text{kN}\,\text{m}^{-2}$. The values of Nt, Mf, Nf and Qf, therefore, need to be multiplied by 5.374.

(c) NB Whatever the following calculations give for shell reinforcement, the minimum shell reinforcement should not be less than, say, 0.2 per cent of the gross cross-sectional area (see Section 2.6). Where the shell is 130 mm thick this is

$$0.002 \times 130 \times 1000 = 260\,\text{mm}^2\,\text{m}^{-1}$$

and where 290 mm thick

$$0.002 \times 290 \times 1000 = 580\,\text{mm}^2\,\text{m}^{-1}$$

(d) The values of Nt, which are for a loading of $1\,\text{kN}\,\text{m}^{-2}$ of curved surface area, are plotted in Fig. 3.1. For $s = 0$–0.45 m, the maximum tension is

$$61.6 \times 5.374 = 331.0\,\text{kN}\,\text{m}^{-1}$$

The area of reinforcement required for this zone can, therefore, be taken as

$$\frac{331\,000}{140} = 2364\,\text{mm}^2\,\text{m}^{-1}$$

say, 25 mm diameter bars at 200 mm centres.

The reinforcement can be assessed similarly for each further 0.45 m increment in s.

The maximum compression is

$$17.8 \times 5.374 = 95.66\,\text{kN}\,\text{m}^{-1}$$

when s is 5.85 or 6.30 m, where the shell is 130 mm thick. The maximum

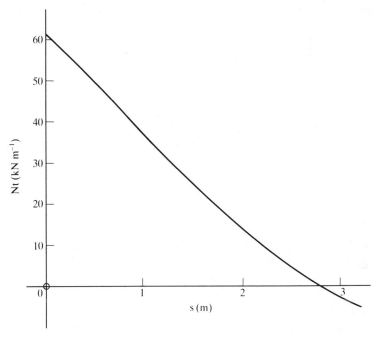

Figure 3.1

compressive stress is therefore

$$\frac{(95\,660/1000)}{130} = 0.736 \,\text{N}\,\text{mm}^{-2}$$

which is much much less than $5.3 \,\text{N}\,\text{mm}^{-2}$.

(e)   The values of Mf, which are for a loading of $1 \,\text{kN}\,\text{m}^{-2}$ of curved surface area, are plotted in Fig. 3.2. The maximum negative bending moment at $s = 0$ is

$$2.02 \times 5.374 = 10.86 \,\text{kNm}\,\text{m}^{-1}$$

At this position the overall depth of shell, h2, equals 290 mm. Assuming waterproofing protection, a cover of 15 mm (see Section 2.5) and 8 mm diameter bars, the effective depth of the reinforcement is

$$290 - 15 - (8/2) = 271 \,\text{mm}$$

Therefore, the moment of resistance to the maximum negative, which

33

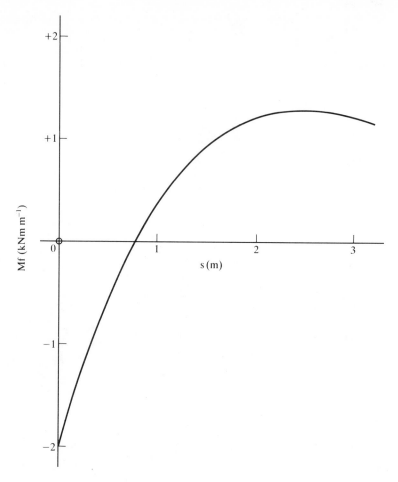

Figure 3.2

is the maximum bending moment is given by

$$\text{moment of resistance} = 1286 \times 1 \times 0.271$$

$$= 94.45 \text{ kNm m}^{-1}$$

which is much greater than 10.86 and is, therefore, satisfactory.

The area of reinforcement required to resist the maximum negative bending moment is

$$\frac{10.86}{140\,000 \times 0.8571 \times 0.271} \times 10^6 = 334 \text{ mm}^2$$

say 8 mm diameter bars at 150 centres. The maximum positive bending moment from Fig. 3.2 at s = 2.45 m is

$$1.28 \times 5.374 = 6.879 \, \text{kNm m}^{-1}$$

At this position the overall depth of shell equals

$$130 + (290 - 130) \times (5.18 - 2.45)/5.18$$

$$= 214 \, \text{mm}$$

Assuming waterproofing protection, a cover of 15 mm (see Section 2.5) and 8 mm diameter bars, the effective depth of the reinforcement is $214 - 15 - (8/2) = 195$ mm. Therefore, the area of reinforcement required equals

$$\frac{6.879}{140\,000 \times 0.8571 \times 0.195} \times 10^6 = 294 \, \text{mm}^2$$

say 8 mm diameter bars at 150 centres. Because of the requirement of (c) previously, it is desirable to use 8 mm diameter bars at 150 centres, top and bottom, in the thickened zone and 8 mm diameter bars at 150 centres in the bottom only in the 130 mm thick zone plus the following amount over the crown. From Table 33A between s = 7.2 m and the crown, the maximum bending moment causing tension in the top of the shell is

$$0.06 \times 5.374 = 0.322 \, \text{kNm m}^{-1}$$

At this position the overall depth of shell equals 130 mm. Assuming waterproofing protection, a cover of 15 mm (see Section 2.5) and 6 mm diameter bars, the effective depth of the reinforcement is

$$130 - 15 - (6/2) = 112 \, \text{mm}$$

Therefore

$$\text{area of reinforcement required} = \frac{0.322}{140\,000 \times 0.8571 \times 0.112} \times 10^6$$

$$= 24 \, \text{mm}^2$$

Hardly any steel is thus required here, use, say, 6 mm diameter bars at 300 mm centres.

The moment of resistance should be checked as was done at s = 0 where the shell is thinner but subjected to smaller bending moments.

(f)   The values of Nf, which are for a loading of 1 kN m^{-2} of curved surface area at s = 4.05 m

$$\text{maximum value} = 17.51 \times 5.374$$

$$= 94.1 \text{ kN m}^{-1}$$

At this position the overall depth of shell is slightly more than 130 mm. Therefore

$$\text{maximum compressive stress} = \frac{94.1}{0.130 \text{ kN m}^{-2}}$$

$$= 0.724 \text{ N mm}^{-2}$$

which is much less than 5.3 N mm^{-2}.

(g) The values of Qf, which are for a loading of 1 kN m^{-1} of curved surface area, and at s = 0

$$\text{maximum value} = 3.4 \times 5.374$$

$$= 18.27 \text{ kN m}^{-1}$$

At this position the effective depth is 271 mm. Therefore

$$\text{maximum shear stress} = \left(\frac{18.27}{0.8571 \times 0.271}\right) \bigg/ 1000$$

$$= 0.0787 \text{ N mm}^{-2}$$

which is much less than 0.7 N mm^{-2} and is, therefore, satisfactory.

(h) The values of ft and fb for the ring beam, are for a loading of 1 kN m^{-2} of curved surface area. The top fibre stress is, therefore, a tensile stress of

$$0.21 \times 5.374 = 1.129 \text{ N mm}^{-2}$$

and the bottom fibre stress is a tensile stress of

$$0.435 \times 5.374 = 2.338 \text{ N mm}^{-2}$$

These are shown in Fig. 3.3. To determine the reinforcement, the beam can be considered in portions, each, say, 300 mm deep as shown in Fig. 3.3. For the bottom portion

$$\text{average tensile stress} = 1.129 + (2.338 - 1.129) \times (7/8)$$

$$= 2.187 \text{ N mm}^{-2}$$

and

$$\text{tensile force} = 2.187 \times 300 \times 300 = 196\,800 \text{ N}$$

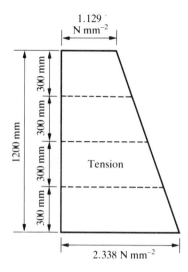

1.129 N mm^{-2}

1200 mm

300 mm 300 mm 300 mm 300 mm

Tension

2.338 N mm^{-2}

Figure 3.3

Thus

$$\text{area of reinforcement required} = \frac{196\,800}{140} = 1406 \text{ mm}^2$$

Use, say, seven 16 mm diameter bars, three in the bottom and one at each side about half-way up the zone, and one at each side at the top of the zone. The reinforcement can be designed similarly for the other portions of the beam.

It is of interest, with regard to surrounding building constructions, to assess the outward movement of the bottom of the ring beam. The increase in the plan radius (i.e. the half span) at the level of the bottom of the ring beam divided by this radius, is the same as the circumferential strain in the bottom fibres of the ring beam. Therefore, using $E = 13\,333$ N mm^{-2} (Section 2.3) the increase in the plan radius (i.e. the outward movement of the bottom of the ring beam) equals

$$\frac{(2.338/13\,333) \times 30}{2} \times 10^3 = 2.63 \text{ mm}$$

This is very small and most manageable.

Example 3.2   Estimate the cost of the dome of Example 3.1.

For estimating the cost, Table 33A gives the values of the curved surface area of the shell soffit, A, as 751 m^2 and the volume of the shell plus its thickening, Vol., as 132.7 m^3.

The cost per m² of the curved shuttering (formwork) to the soffit of the shell is multiplied by 751 m².

The concrete price per m² for the shell is built up of the cost per m² for finishing (tamping, trowelling, etc.) and the cost of providing its volume to its location. This latter requires knowledge of the volume of concrete per m² of shell and an average figure for this is

$$\frac{132.7}{751} = 0.1767 \, \text{m}^3 \, \text{m}^{-2}$$

This is a normal way for estimating the cost of shells when working for designer-contracting firms. (Alternatively, some estimators might price the concrete per m³ (i.e. using 132.7 m³), and just bear in mind in their costing that the average thickness of the shell is 132.7/751 m = 177 mm.)

Example 3.3 Design a dome, between tables with different ring beam sizes, of plan diameter (span) 17.8 m and of type B, shown in Fig. 2.2, to carry its self-weight and a loading of 1.205 kN m⁻² (see Section 2.1).

Refer to Tables 14B where DIA. = 17.6 m and 15B where DIA. = 18.0 m, to Section 2.7, and Figs 2.4, 2.5 and 2.6. Between these tables the size of the ring beam has been altered. This makes a considerable difference in the values, for example, of Mf. One can, therefore, use either Tables 13B and 14B, or 15B and 16B, dependent on which is nearest to the value of DIA. and bearing in mind that it is more prudent to choose the larger ring beam. In this example the latter is chosen. Values from Tables 15B and 16B are obtained by extrapolation, or taking the larger value if close enough together, as can be seen in Example 3.4. Alternatively, some values can easily be obtained by direct calculation, as shown in Example 3.6.

3.2 Non-prestressed Domes, USA/British Imperial Units (or SI Units between Tables)

The following example shows how to deal with any dome which is of a span between two of the design tables, whether, USA/British Imperial or SI units.

The elastic theory and stresses, etc. as given in Section 3.1.1 will be used in the example.

Example 3.4 Design a dome of plan diameter (span) 95 ft 5 in (29.083 m) and of type C, shown in Fig. 2.3, to carry its self-weight and a loading of 25.17 psf (1.205 kN m⁻²) (see Section 2.2).

Use Tables 31C where DIA. = 28.7 m and 32C where DIA. = 29.5 m, and refer to Section 2.7 and Figs 2.4, 2.5 and 2.6. By interpolation these tables can be used as follows:

1 DIA. = 29.083 m (95 ft 5 in), restated
2 Using Table 31C

$$\text{RISE} = 3.578 \times \frac{29.083}{28.7}$$

$$= 3.626 \text{ m } (11 \text{ ft } 10.8 \text{ in})$$

or using Table 32C

$$\text{RISE} = 3.678 \times \frac{29.083}{29.5}$$

$$= 3.626 \text{ m } (11 \text{ ft } 10.8 \text{ in})$$

3 Using Table 31C

$$\text{RADIUS} = 30.563 \times \frac{29.083}{28.7}$$

$$= 30.971 \text{ m } (101 \text{ ft } 7.3 \text{ in})$$

or using Table 32C

$$\text{RADIUS} = 31.415 \times \frac{29.083}{29.5}$$

$$= 30.971 \text{ m } (101 \text{ ft } 7.3 \text{ in})$$

4 Using Tables 31C and 32C, ALPHA = 28°, h1 = 100 mm (3.9 in), h2 = 250 mm (9.8 in)
5 Using Table 31C, (notice the square root)

$$S = 4.625 \times \sqrt{(29.083/28.7)}$$

$$= 4.656 \text{ m } (15 \text{ ft } 3.3 \text{ in})$$

or using Table 31C

$$S = 4.689 \times \sqrt{(29.083/29.5)}$$

$$= 4.656 \text{ m } (15 \text{ ft } 3.3 \text{ in})$$

6 Using Tables 31C and 32C, b = 300 mm (11.8 in), d = 1000 mm (3 ft 3.4 in), y = −443 mm (1 ft 5.4 in), db = 0, qd = 2.65 kN m^{-2} (in this example, but qd varies in a complex way with span)
7 Using Table 31C (notice the inverse square)

$$qb = 4.28 \times (28.7/29.083)^2$$

$$= 4.168 \text{ kN m}^{-2} \text{ (87.05 psf)}$$

or using Table 32C,

$$qb = 4.05 \times (29.5/29.083)^2$$
$$= 4.167 \text{ kN m}^{-2} \text{ (87.03 psf)}$$

8   Using Table 31C

$$p = 29.875 \times (29.083/28.7)$$
$$= 30.274 \text{ m (99 ft 3.9 in)}$$

or using Table 32C

$$p = 30.708 \times (29.083/29.5)$$
$$= 30.274 \text{ m (99 ft 3.9 in)}$$

9   Using Table 31C (notice the square)

$$A = 687 \times (29.083/28.7)^2$$
$$= 705 \text{ m}^2 \text{ (7593 ft}^2)$$

or using Table 32C

$$A = 726 \times (29.083/29.5)^2$$
$$= 706 \text{ m}^2 \text{ (7595 ft}^2)$$

The minute difference is due to the rounding off of these quantities on the tables to the nearest square metre, giving a maximum possible error in this case of, say, $100 \times (0.5/700) = 0.07$ per cent which may be regarded as negligible.

10 qD has a complex algebraic and trigonometrical relationship with the span. The tables are quite close together in that Table 31C gives $qD = 3.331 \text{ kN m}^{-2}$, whilst Table 32C gives $qD = 3.319 \text{ kN m}^{-2}$. The difference is only 0.36 per cent so one could simply use the larger quantity. Otherwise a linear interpolation, for what it is worth, could be used giving

$$3.331 - \frac{(3.331 - 3.319) \times (29.083 - 28.7)}{(29.5 - 28.7)} = 3.325 \text{ kN m}^{-2}$$

11 ft has a complex algebraic and trigonometrical relationship with the span. The tables are quite close together in that Table 31C gives $ft = -0.221 \text{ kN m}^{-2}$, whilst Table 31C gives $ft = -0.219 \text{ kN m}^{-2}$. The difference is only 0.91 per cent so one could simply use the larger quantity. Otherwise a linear interpolation, for what it is worth, could be used giving

$$-0.221 + \frac{(0.221 - 0.219) \times (29.083 - 28.7)}{(29.5 - 28.7)} = -0.220 \text{ N mm}^{-2}$$

12   fb has a complex algebraic and trigonometrical relationship with the span. The tables are quite close together in that Table 31C gives fb $= 0.764$ kN m^{-2}, whilst Table 32C gives fb $= 0.796$ kN m^{-2}. The difference is only 4.19 per cent so one could simply use the larger quantity. Otherwise a linear interpolation, for what it is worth, could be used giving

$$0.764 + \frac{(0.796 - 0.764) \times (29.083 - 28.7)}{(29.5 - 28.7)} = 0.779 \text{ N mm}^{-1}$$

13   The volume Vol. has a complex algebraic and trigonometrical relationship with the span. The tables are quite close together in that Table 31C gives Vol. $= 97.0$ m^3, whilst Table 32C gives Vol. $= 102.1$ m^3. The difference is only 5.3 per cent so one could simply use the larger quantity. Otherwise a linear interpolation, for what it is worth, could be used giving

$$97.0 - \frac{(102.1 - 97.0) \times (29.083 - 28.7)}{(29.5 - 28\text{-}7)} = 94.6 \text{ m}^3 \ (3341 \text{ ft}^3)$$

14   In both Tables 31C and 32C the increments of s are the same from 0 to 7.20 m. Beyond this does not matter to the design. In any case the tables are close enough together to just simply take the larger quantities, that is, to use the values of Nt, Mf, Nf and Qf given in Table 32C for all the various values of s given there. For example, for s $= 0$ the values of Nt are 147.9 and 154.2, so taking the 154.2 as opposed to an interpolated value of

$$147.9 + \frac{(154.2 - 147.9) \times (29.083 - 28.7)}{(29.5 - 28.7)} = 150.9$$

one is only 2.18 per cent overgenerous. Otherwise interpolated values can be used.

All these values of Nt, Mf, Nf and Qf are for a loading of 1 kN m^{-2} (not unit psf or ksf, so it is important to work in kN m units and if one's design is in traditional USA/British Imperial units one needs to convert the final forces and moments from kN m units to pound, feet or inch units) of curved surface area and they must be multiplied by the total loading in kN m^{-2}. In this example

$$\text{total loading} = 1.205 + qD$$
$$= 1.205 + 3.325$$
$$= 4.53 \text{ kN m}^{-2}$$

So for example, from the previous paragraph, at s = 0

$$Nt = 150.9 \times 4.53 = 683.6 \, \text{kN m}^{-1}$$

The design is then conducted similarly to that in Example 3.1.

(a) The self-weight of the shell and 30 per cent of its thickenings (see Section 2.4, and point 6 previously) equals

$$qd = 2.65 \, \text{kN m}^{-2}$$

Adding this to 1.205 for this total loading gives a buckling load of 3.855 kN m^{-2}. This is less than the safe buckling loading (see Section 2.3 and point 7, previously) i.e. qb = 4.167 kN m^{-2}, and is therefore satisfactory.

(b) The self-weight of the shell and of its thickenings (see Section 2.4, and point 10 previously) equals qD = 3.325 kN m^{-2}. Adding this to 1.205 gives 4.53 kN m^{-2}. The values of Nt, Mf, Nf and Qf, therefore, need to be multiplied by 4.53.

(c) NB Whatever the following calculations give for shell reinforcement the minimum shell reinforcement should not be less than, say, 0.2 per cent of the gross cross-sectional area (see Section 2.6). Where the shell is 100 mm (3.9 in) thick this is

$$0.002 \times 100 \times 1000 = 200 \, \text{mm}^2 \, \text{m}^{-1}$$

$$(0.002 \times 3.9 \times 12 = 0.936 \, \text{in}^2/\text{ft})$$

and where 250 mm (9.8 in) thick

$$0.002 \times 250 \times 1000 = 500 \, \text{mm}^2 \, \text{m}^{-1}$$

$$(0.002 \times 9.8 \times 12 = 0.235 \, \text{in}^2/\text{ft})$$

(d) The values of Nt, which are for a loading of unit kN m^{-2} of curved surface area, and are dealt with in point 14 previously, can be plotted similarly to Fig. 3.1. These have to be multiplied by 4.53 kN m^{-2} to obtain values of N_θ. The reinforcement required to resist the tensile stresses is designed similarly to Example 3.1(d). The maximum compressive stress is checked as in Example 3.1(d).

(e) The values of Mf, which are for a loading of unit kN m^{-2} of curved surface area and are dealt with in point 14 previously, can be plotted similarly to Fig. 3.2. These have to be multiplied by 4.53 kN m^{-2} to obtain values of M_ϕ. The maximum extreme fibre compressive stresses are checked as in Example 3.1(e). The reinforcement required to resist these bending moments is designed as in Example 3.1(e).

(f) The values of Nf, which are for a loading of unit kN m^{-2} of curved surface area, and are dealt with in point 14 previously, have to be multiplied by $4.53 \, \text{kN} \, \text{m}^{-2}$ to obtain values of N_ϕ. The maximum compressive stresses are checked as in Example 3.1(f).

(g) The values of Qf, which are for a loading of unit $\text{kN} \, \text{m}^{-2}$ of curved surface area, and are dealt with in point 14 previously, have to be multiplied by $4.53 \, \text{kN} \, \text{m}^{-2}$ to obtain values of Q_ϕ. The maximum shear stresses are checked similarly to Example 3.1(g).

(h) The values of ft and fb for the ring beam, which are for a loading of unit $\text{kN} \, \text{m}^{-2}$ of curved surface area, and are dealt with in points 11 and 12 previously, have to be multiplied by $4.53 \, \text{kN} \, \text{m}^{-2}$ to obtain values of f_t and f_b. These can be plotted as in Fig. 3.3. The reinforcement required to resist the tensile stresses is designed as in Example 3.1(h).

In this instance because the ring beam is upstand, the top fibre stress is compressive and equals

$$-0.220 \times 4.53 = -0.997 \, \text{N} \, \text{mm}^{-2} \, (-144.5 \, \text{psi})$$

Although this ring beam is curved, which is helpful in lateral restraint, it is only 300 mm wide and its length is

$$\pi \times (29.083 + 0.3) = 29.48 \, \text{m}$$

ACI 318–89 requires the spacing of lateral restraints to be not greater than $50b$; BS 8110 recommends not greater than the lesser of $60b$ or $250b/(\text{effective depth required})$ whilst $30b$ is regarded as ideal, between $30b$ and $40b$ as reasonable, and more than $50b$ likely to be impracticable.[3.1] The ratio of length to breadth of the ring beam equals $29.48/0.3 = 98.3$, so stiffeners to prevent lateral instability are required. As the structure is particularly three-dimensional the minimum number of stiffeners can be regarded as three, giving the spacing of lateral restraints as $98.3b/3 = 32.8b$, which can be regarded as satisfactory. These stiffeners can be struts of 100 mm by 100 mm (4 in by 4 in) cross-section with, say, four 12 mm (1/2 in) diameter bars longitudinally, one in each corner and 6 mm (1/4 in) diameter stirrups at 150 mm (6 in) centres. Struts of this type are shown in Fig. 2.3.

It is of interest, with regard to surrounding building constructions, to assess the outward movement of the bottom of the ring beam. The increase in the plan radius (i.e. the half span) at the level of the bottom of the ring beam divided by this radius is the same as the circumferential strain in the bottom fibres of the ring beam. The stress in these latter is

$$0.779 \times 4.53 = 3.529 \, \text{N} \, \text{mm}^{-2}.$$

Therefore, using $E_c = 13\,333 \, \text{N} \, \text{mm}^{-2}$ (from Section 2.3) the increase in the plan radius (i.e. the outward movement of the bottom of the ring beam) equals

$$\frac{(3.529/13\,333) \times 29.083}{2} \times 10^3 = 7.70 \, \text{mm} \, (0.30 \, \text{in})$$

This is small and manageable.

Example 3.5 Estimate the cost of the dome of Example 3.4.

For estimating the cost, Example 3.4(9) and (13) give the values of A as 706 m² (7593 ft²) and Vol. as 94.6 m³ (3341 ft³) respectively.

The cost per m² (or ft²) of the curved shuttering (formwork) to the soffit of the shell is multiplied by 705 m² (or 7593 ft²).

The concrete price per m³ (or ft³) for the shell is built up of the cost per m³ (or ft³) for finishing (tamping, trowelling, etc.) and the cost of providing its volume to its location. This latter requires knowledge of the volume of concrete per m³ (or ft³) of shell, and an average figure for this is

$$\frac{94.6}{705} = 0.1342 \, \text{m}^3 \, \text{m}^{-2}, \quad \text{or} \quad \frac{3341}{7593} = 0.4400 \, \text{ft}^3/\text{ft}^2$$

This is one way of estimating the cost of shells when working for designer-contracting firms.

Alternatively some estimators might price the concrete per m³ or ft³ (i.e. using 94.6 m³, or 3341 ft³) and just bear in mind in their costing that the average thickness of the shell is

$$\frac{94.6}{705} = 0.1342 \, \text{m} = 134 \, \text{mm} \, (\text{or} \, 5.28 \, \text{in})$$

Example 3.6 Direct calculation for some of the quantities in Examples 3.4 and 3.5.

Alternatively to some of the values obtained in Example 3.4 and 3.5, the following can be obtained fairly easily by direct calculation:

DIA. = 29.083 m (95 ft 5 in)

ALPHA = 28°

RADIUS = DIA./2/sin α = 29.083/2/sin 28°

$\qquad$ = 30.974 m (101 ft 7.449 in)

RISE = RADIUS $\times$ (1 $-$ cos α) = 30.974 $\times$ (1 $-$ cos 28°)

$\qquad$ = 3.626 m (11 ft 10.739 in)

p = 2α $\times$ RADIUS = 2 $\times$ (28 $\times$ π/180) $\times$ 30.974

$\qquad$ = 30.274 m (99 ft 3.870 in)

A = 2 $\times$ π $\times$ RADIUS² $\times$ (1 $-$ cos α)

$\qquad$ = 2 $\times$ π $\times$ 16.469² $\times$ (1 $-$ cos 28°)

$\qquad$ = 199.48 m² (2147.2 ft²)

3.3 Prestressed Domes, SI Units

In the following examples prestressing wires, or strands, post-tensioned around the outer face of the ring beam are considered. It is shown how the prestressing force is calculated. A variety of wires and strands are available and a selection should be made in relation to whatever code of practice is followed. In the examples tentative suggestions are made to indicate the practicability of accommodating tendons within the depth of the ring beams.

Wires can be tensioned around the ring beams with well-known machines originally developed for winding and tensioning wires around prestressed circular tanks, and the ring beams above them, when the roofs were domes (see Fig. 3 of Ref. 3.1).

Drawn-strands, (these are smoother than normal strands) supplied covered in grease, the whole contained in a plastic sheath, may be used to externally prestress the ring beam of a dome that has inadequate ring tension reinforcement due to detailing errors. These may be anchored as shown in Fig. 3.4. In a practical example undertaken by the author, to reduce friction, junctions were made at 1/8 points around the periphery and each drawn-strand was jacked from both of its ends. When jacked, each strand was found to be perfectly mobile, the slightest overpressure in either jack causing the whole strand to immediately move that way. Had it been known that the lubrication would be so fantastically effective, the number of junctions around the periphery may have been reduced. However, friction between such tendons and the concrete is difficult to assess with precision.

Example 3.7 Design the dome of Example 3.1, but with a prestressed ring beam.

From Example 3.1 the maximum value of N_θ is at $s = 0$ and is $331.0 \, \mathrm{kN \, m^{-1}}$. This tension can be eliminated by prestressing. Use Table 33AP. For a prestressing force of $1000 \, \mathrm{kN}$ this gives at $s = 0$, a value of $Nt = -333.7 \, \mathrm{kN \, m^{-1}}$. Therefore, the prestressing force required is

$$1000 \times (331/333.7) = 991.9 \, \mathrm{kN}$$

If for example, $18 \, \mathrm{mm}$ $(0.7 \, \mathrm{in})$ nominal diameter drawn-strand (see Section 3.3 and Fig. 8.4 in Ref. 3.2) is used, having a force after losses of $200 \, \mathrm{kN}$, then the number of these required is $991.9/200 = 4.96$, say 5. Their centre of pressure must be at the height of the half-depth of the ring beam.

Then the values of Nt from Table 33AP times 0.9919 are superimposed upon the values from Table 33A of Nt times 5.374 to obtain resultant values of N_θ. If, because of the different distributions of Nt on the two tables, a tension happens to occur at some value of s then the prestressing force can be increased so that nowhere is the resultant N_θ in tension.

The prestressing force decided upon is then multiplied by the values of

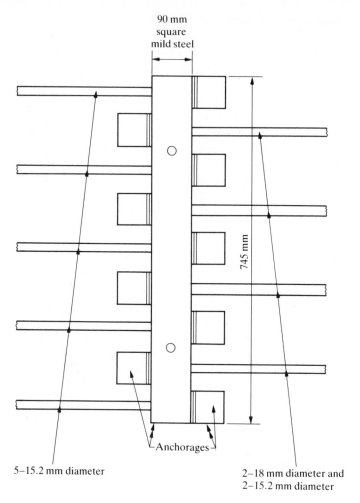

90 mm
square
mild steel

745 mm

Anchorages

5–15.2 mm diameter

2–18 mm diameter and
2–15.2 mm diameter

Figure 3.4

Mf, Nf and Qf from Table 33AP and these are superimposed upon the values of these quantities from Table 33A multiplied by 5.374. The design for M_ϕ, N_ϕ and Q_ϕ is then similar to that of Example 3.1, as is the assessment of the horizontal movement of the bottom of the ring beam.

The top fibre stress in the ring beam from Example 3.1 was 1.129 N mm^{-2}. The top fibre stress due to prestressing, from Table 33AP is

$$-1.013 \times 0.9919 = -1.005 \text{ N mm}^{-2}$$

Therefore, the resultant top fibre stress is

$$1.129 - 1.005 = 0.124 \, \text{N} \, \text{mm}^{-2} \, (\text{c} \, 18 \, \text{psi})$$

which is negligible, but if one wants to make it compressive then the prestressing force can be increased by a minute amount.

The bottom fibre stress in the ring beam from Example 3.1 was $2.338 \, \text{N} \, \text{mm}^{-2}$. The bottom fibre stress due to prestressing, from Table 33AP is

$$-2.786 \times 0.9919 = -2.763 \, \text{N} \, \text{mm}^{-2}$$

Therefore, the resultant top fibre stress equals

$$2.338 - 2.763 = -0.425 \, \text{N} \, \text{mm}^{-2} \, (\text{c} \, 62 \, \text{psi})$$

which is negligible but compressive and, therefore, satisfactory.

The outward movement of the bottom of the ring beam is obtained from the resultant bottom fibre stress as explained in Example 3.1. Therefore, in this present case, movement is inwards and is

$$\frac{(0.425/13\,333) \times 30}{2} \times 10^3 = 0.48 \, \text{mm}$$

This is negligible and most manageable.

One may also be interested in stresses and the horizontal movement of the bottom of the ring beam when the prestressing force is in place but the dome carries no superimposed loading, which in some instances can be for most of its life. This is ascertained as above; the loading of 5.374 is reduced by (see Section 2.2) the superimposed loading of $0.75 \, \text{kN} \, \text{m}^{-2}$.

Example 3.8 Estimate the cost of the dome of Example 3.7.

This is the same as in Example 3.2 except for the added expense of supplying and post-tensioning five 18 mm diameter drawn-strands in greased plastic sheaths (see Example 3.7).

Example 3.9 Design the dome of Example 3.3 with a prestressed ring beam.

The procedure is similar to that of Example 3.3. As in that example, the larger ring beam is chosen and if values are obtained by extrapolation from Tables 15B and 16B then values will be extrapolated from Tables 15BP and 16BP to consider the effect of the 1000 kN prestressing force. The actual prestressing force required is then obtained and the design continued as in Example 3.7.

3.4 Prestressed Domes, USA/British Imperial Units or SI Units between Tables

The following example shows how to deal with any dome which is of a span between two of the design tables whether USA/British Imperial or SI units.

The elastic theory and stresses, etc. as given in Section 3.1.1 will be used in the example.

Example 3.10   Design the dome of Example 3.4 with a prestressed ring beam.

From Example 3.4(14) the maximum value of N_θ is at s = 0 and is, using the interpolated value, 683.6 kN m^{-1}.

This tension can be eliminated by prestressing. Tables 31CP and 32CP give, at s = 0, for a prestressing force of 1000 kN, values of Nt of -298.4 and -299.1 kN m^{-1} respectively. By interpolation

$$Nt = -298.4 + \frac{(299.1 - 298.4) \times (29.083 - 28.7)}{(29.5 - 28.7)} = -298.7 \text{ kN m}^{-1}$$

Therefore the prestressing force required is

$$\frac{1000 \times 683.6}{298.7} = 2289 \text{ kN (514.6 kips)}$$

If for example, 7 mm (0.276 in) diameter wires (see Section 3.3 and Fig. 8.4 in Ref. 3.2) are used with a stress after losses of 920 N mm^{-2} (130.5 ksi) then the area of these required is

$$2289/0.93 = 2461 \text{ mm}^2 \text{ (3.81 in}^2)$$

and the number required is

$$2461/38.48 = 63.96, \text{ say } 64.$$

These can be spread out on the outside of the ring beam of depth 1000 mm (3.281 ft). Their centre of pressure must be at the height of the half-depth of the ring beam. They could be spaced at, say, 15 mm centres taking up a space of

$$(64 - 1) \times 15 = 945 \text{ mm (37.2 in)}$$

Had it been decided in Example 3.4 to take the larger quantities of Table 32C, because of their nearness to those of Table 31C, then for this present case with the prestressed ring beam, the quantities of Table 32CP would be used as these are also very near to the values of Table 31CP.

Then the values of Nt by interpolation between Tables 31CP and 32CP (or 32CP if the immediately previous paragraph is selected) times 2.289 are superimposed upon the values of N_θ from Example 3.4(14) to obtain resultant values of N_θ. If, because of the different distributions of Nt on the two tables, a tension happens to occur at some value of s then the prestressing force can be increased so that nowhere is the resultant N_θ in tension.

The prestressing force decided upon is then multiplied by the values of Mf, Nf and Qf from Table 32CP (or from interpolation between Tables 31CP and 32CP, if the design had used this preference) and these are superimposed upon the values of M_ϕ, N_ϕ and Q_ϕ from Example 3.4(14).

The design for the resultant values of M_θ, N_ϕ and Q_ϕ is then similar to that of Example 3.1, as is the assessment of the horizontal movement of the bottom of the ring beam.

The top fibre stress in the ring beam from Example 3.4 was $-0.997\,\mathrm{N\,mm^{-2}}$ ($-144.5\,\mathrm{psi}$). The top fibre stress due to prestressing, from interpolation between Tables 31CP and 32CP is

$$-\left[1.403 - \frac{(1.403 - 1.377) \times (29.083 - 28.7)}{(29.5 - 28.7)}\right] \times 2.289$$

$$= -(1.403 - 0.012) \times 2.289$$

$$= -1.391 \times 2.289$$

$$= -3.184\ \mathrm{N\,mm^{-2}}$$

Therefore the resultant top fibre stress is

$$-0.997 - 3.184 = -4.181\ \mathrm{N\,mm^{-2}}\ (606.4\ \mathrm{psi})$$

which is compressive and satisfactory.

Lateral instability can be dealt with by providing struts as in Example 3.4.

The bottom fibre stress in the ring beam from Example 3.4 was $0.779 \times 4.53 = 3.529\ \mathrm{N\,mm^{-2}}$ ($511.8\,\mathrm{psi}$). The bottom fibre stress due to prestressing from interpolation between Tables 31CP and 32CP is

$$-\left[1.476 + \frac{(1.48 - 1.476) \times (29.083 - 28.7)}{(29.5 - 28.7)}\right] \times 2.289$$

$$= -3.383\ \mathrm{N\,mm^{-2}}\ (-490.7\ \mathrm{psi})$$

Therefore the resultant bottom fibre stress is

$$3.529 - 3.383 = 0.146\ \mathrm{N\,mm^{-2}}\ (21.2\ \mathrm{psi})$$

which is negligible, but if one wants to make it compressive then the prestressing force can be increased by a minute amount.

One may also be interested in stresses and the horizontal movement of the bottom of the ring beam when the prestressing force is in place, but the dome carries no superimposed loading, which in some instances can be for most of its life. This is calculated as given above; the loading of 4.53 is reduced by (see Section 2.2) the superimposed loading of $0.75\,\mathrm{kN\,m^{-2}}$.

Example 3.11 Estimate the cost of the dome of Example 3.10. This is calculated as in Example 3.2 except for the added expense of supplying and post-tensioning (by, say, winding) sixty-four 7 mm (0.276 in) diameter wires (see Example 3.10).

3.5 Design of Domes Supported with Columns

The domes considered previously in this book have their ring beams supported continuously and uniformly by walls of buildings or tanks. It is not uncommon for buildings to require domes to be supported by columns. Essentially, it is desirable to have these closely spaced, relative to the depth of the ring beam, to approximate to the conditions assumed in the shell analysis. Not to do this means that the boundary conditions between shell and ring beam are interfered with, and if for example this were a severe interference such as in the extreme case of having only three supports, then the benefits of the shell analysis would suffer. It would be necessary to consider bending moments, with associated shear forces, at right angles to the M_ϕ moments. These would be of more significance near to the ring beam than near the crown.

Many papers and books contain photographs and sketches showing domes supported by columns; however, none could actually be used by a designer to design practical domes supported by columns. For example, one book says 'shell concrete domes can be supported on three or more columns', then gives no more information on how to design domes supported by columns and only gives the membrane formulae of Eqs 1.1 and 1.2; and does not even suggest matching up boundary conditions between shell and ring beam.

Timoshenko[3.3] has a section 'spherical shell supported at isolated points'. Essentially he represents the continuous load from the shell edge, combined with the point support loads, as a mathematical series. He then limits coverage to the membrane theory, and says, 'so long as we limit ourselves to membrane theory only, we shall not have enough constants to satisfy all the conditions and to obtain the complete solution of the problem'. Timoshenko then suggests that, if a ring beam is used, his mathematical series can be used with Eqs 1.1 and 1.2; but this only considers membrane forces in the shell and does not deal with the practical desirability of matching up boundary conditions between shell and ring beam.

In the early days of shells in the UK the edge beams of cylindrical shell roofs were often assumed to be continuously supported in the analysis, and to approach this assumption columns were provided preferably at not more than 15 ft (4.57 m) centres. When the columns were closer $\leqslant 10$ ft (3.05 m) apart the edge beam was made 30 in (762 mm) deep and used nominal longitudinal steel (c 7.5 lb/ft run, i.e. 0.1094 kN m^{-1}, i.e. an area of 1421 mm^2).

When further apart than this the edge beam was designed as a normal continuous beam carrying about one-eighth of the chord width of the adjacent shell and its loading. On this approximate basis one perhaps could risk applying the same crude idea to columns beneath ring beams, using the dome span as the chord width; it is up to the designer!

In simple approximate terms, for modest spacings of columns, the ring beam and its adjacent shell, must be strong enough to span continuously between the columns. It would seem sensible to count upon the strength and stiffness of the full depth of the ring beam. Then the stresses are reduced according to how much depth of the shell is also counted upon. The following example uses this approximate design method.

Example 3.12   Design the dome of Example 3.1 to be supported on columns.

Suppose there are twenty columns beneath the ring beam, the length of the ring beam is

$$\pi \times 30.3 = 95.190 \text{ m}$$

The spacing of the columns is

$$95.190/20 = 4.760 \text{ m}$$

From Example 3.1(b) the total shell loading is

$$5.374 \text{ kN m}^{-2}$$

From Table 33A, $A = 751 \text{ m}^2$. Therefore, the total load from the shell on to the ring beam is

$$5.374 \times 751 = 4036 \text{ kN} \qquad \centerdot$$

which per unit length of ring beam is

$$4036/95.190 = 42.40 \text{ kN m}^{-1}$$

The weight of the ring beam per unit length (taking the weight density of concrete as 23.6 kN m^{-3}) is

$$0.3 \times 1.2 \times 23.6 = 8.50 \text{ kN m}^{-1}$$

Therefore, the supporting force necessary is

$$42.40 + 8.50 = 50.9 \text{ kN m}^{-1}$$

The ring beam and part of the shell, according to the designer's judgement, can be used to carry this loading between the columns in a continuous fashion. A support moment will be

$$\frac{50.9 \times 4.76^2}{12} = 96.11 \text{ kN m}$$

and a mid-span moment will be

$$96.11/2 = 48.05 \, \text{kN m}$$

To be on the safe side it may be that the ring beam without the help of the adjacent shell can be reasonably well designed adequately to carry this loading.

Some designers may effect this design, as for a reinforced concrete beam according to the elastic theory cracked in tension, and just detail the reinforcement to this continuous beam and put it in as an extra to its reinforcement designed as in Example 3.1.

It might be considered that this is a little illogical. Most assumptions generally introduce illogicalities. However, this is in line with the methods of Hillberborg,[3.2] where to put it very simply, he makes sure that all loading is adequately supported somehow or other without worrying about matching up displacements. It is, of course, an ultimate load method which assumes that an overloaded part can hold on plastically until other parts reach their maximum carrying capacities.

However, it is recommended that a logical approach be adopted in keeping with the elastic method used in Example 3.1, and analysing the ring beam to span between columns as though uncracked; then superimposing these stresses upon the stresses obtained in Example 3.1, and then designing the reinforcement to take the tensile stresses.

The section modulus of the edge beam equals

$$\frac{(0.3 \times 1.2)^2}{6} = 0.072 \, \text{m}^3$$

At the column support the top fibre stress, for just spanning between columns, is

$$96.11/0.072 = 1335 \, \text{kN m}^{-2}$$
$$= 1.335 \, \text{N mm}^{-2}$$

and the bottom fibre stress is

$$-1.335 \, \text{N mm}^{-2}$$

At mid-span the top fibre stress equals

$$1.335/2 = -0.667 \, \text{N mm}^{-2}$$

and the bottom fibre stress is

$$0.667 \, \text{N mm}^{-2}$$

From Example 3.1, therefore, the resultant stresses are:

(a) at the support:

top fibre stress

$$1.129 + 1.335 = 2.464 \ N \ mm^{-2}$$

bottom fibre stress

$$2.338 - 1.335 = 1.003 \ N \ mm^{-2}$$

(b) at the mid-span:

top fibre stress

$$1.129 - 0.667 = 0.462 \ N \ mm^{-2}$$

bottom fibre stress

$$2.338 + 0.667 = 3.005 \ N \ mm^{-2}$$

These are shown in Fig. 3.5. None of these stresses are compressive, so the compressive strength of the concrete does not need to be examined. It is not advised to reduce the reinforcement of Example 3.1 in any way, because of, say, using Fig. 3.5 in lieu of Fig. 3.3, since, for example, the bottom fibre stress of $2.338 \ N \ mm^{-2}$ at the column support may not be reduced by 1.335 as some of the shell will add to the section modulus of the ring beam and reduce this stress of $-1.335 \ N \ mm^{-2}$. Instead, the designer should favour providing reinforcement to resist the tensile stresses due to the bending moments in the ring beam, see Fig. 3.6, and adding this to that required in Example 3.1.

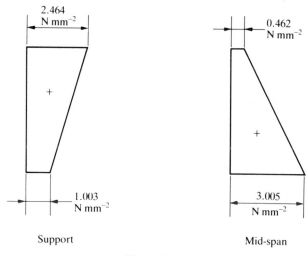

Figure 3.5

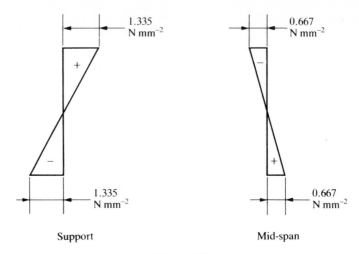

Support   Mid-span

Figure 3.6

In this case, referring to Fig. 3.6, the reinforcement required over the support has to resist a tensile force of

$$\left(\frac{1.335}{2} \times \frac{300 \times 1200}{2}\right)\Big/1000 = 120.2\,\text{kN}$$

The area of reinforcement required for this is

$$120\,200/140 = 859\,\text{mm}^2$$

Two 25 mm diameter bars could be used in the top, or more smaller bars distributed as per Fig. 3.6. The area of reinforcement required at mid-span is therefore

$$859/2 = 429\,\text{mm}^2$$

Two 20 mm diameter bars could be used in the bottom, or more smaller bars distributed as per Fig. 3.6. These are in addition to the reinforcement required in Example 3.1 and a combined reinforcement system will no doubt be devised.

3.6 Domes with Roof Lights

Roof lights are generally not required in domes over tanks containing water, chemicals, sewage and toxic waste. When domes are over buildings, natural lighting is perhaps best effected in the walls beneath the domes, as such windows can be easily equipped with controlled ventilation and blinds which keep out solar heat.

Sometimes roof lights are desired in domes over buildings. An easy way is to use precast concrete panels with small glass lenses, and specialist firms that manufacture these exist. The panels can be made of the same thickness as the shell and the reinforcement in them can be the same as required by the shell at their location.

Circular roof lights of not more than 4 ft (1.219 m) diameter have been constructed in cylindrical shells without altering the shell analysis in any way. These roof lights consist of light plastic, and the openings in the shell have kerbs projected about 150 mm (c 6 in) above the shell surface, and are about 100 mm (c 4 in) wide, see Fig. 3.7. These kerbs are reinforced with nominal steel, for example 6 mm (c 1/4 in) diameter rectangular stirrups (links) with four 12 mm (c 1/2 in) diameter longitudinal bars, one in each corner of the links. These openings are limited to not more than nine per cent of the plan area. A designer might choose to risk this procedure; domes which are doubly curved being stronger in this respect than singly curved cylindrical shells.

In Fig. 3.8 lighting is made by having a substantial opening at the crown bounded by kerbs integral with the shell. The kerbs may support a steel-framed glazing arrangement, perhaps incorporating ventilation arrangements. If the opening is large the boundary conditions have to be solved at the junction between the kerb (which is a ring beam carrying the vertical load from the skylight glazing) and the shell. This analysis has to be carried out simultaneously with the boundary condition problem tackled in Chapter 1 between shell and ring beam at the shell's periphery. This is not done in this book.

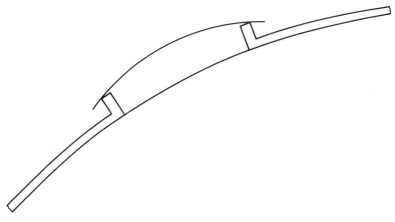

Figure 3.7

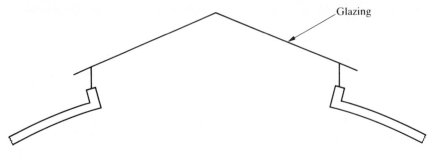

Figure 3.8

If the opening at the crown is not too large, a designer might choose to risk an analogous procedure to that used for cylindrical shells, given as follows. For strip lantern lights along the crown of a cylindrical shell a kerb about 9 in (c 225 mm) high by about 6 in (c 150 mm) wide would be assumed to transmit the forces and moments from the shell (next to the crown) into struts between the kerbs. The struts are at least 9 in (c 225 mm) wide by 9 in (c 225 mm) deep, at not more than 10 ft (3.048 m) centres.

3.7 Temperature Movement and Stresses

The coefficient of linear thermal expansion of structural concrete is about 0.000 01 °C^{-1} (c 0.000 0055 °F^{-1}). The ring beam will, therefore, move outwards and inwards according to the ambient temperature. This latter is a temperature in the shade when the shell is covered with, for example, roofing felt (and thermal insulation helps). If the dome is over, say, an air-conditioned building in the USA, then the maximum temperature is only achieved if the air-conditioning breaks down. In winter a building is usually occupied and, therefore, has some heating. Should this breakdown be coincidental with a very cold spell, it takes some time for the fabric of the building to become as cold as outside, and the central heating would be repaired, or some improvisation made, in this time.

It has been shown previously how to calculate the horizontal structural movement of the ring beam, and this has to be allowed for in the building construction details so that the dome movement does not damage walls, etc. Similarly, the temperature movement of the ring beam should be assessed and allowed for in the same way.

Thermal stresses will occur in the dome if it is not allowed to move freely. Consideration of the details of support on walls or columns can approximately allow this freedom.

Differential thermal stresses are commonly ignored in the design of domes for roofs of tanks and buildings. The shells are relatively thin and the thermal conductivity of concrete is fairly high. Also domes for buildings are often covered with thermal insulation. This can, of course, be done for a tank if it held hot liquids (which is uncommon as most hold cool water) to reduce the temperature difference between its inner and outer surfaces. A great number of cylindrical shells to buildings and tanks, designed ignoring differential thermal stresses, exist. Whether or not to ignore these differential thermal stresses is, of course, a decision of the designer for the circumstances as he judges them.

Example 3.13 Determine the thermal movement of the dome of Example 3.1 if in a building.

If in Example 3.1 the concrete is cast at a temperature of about $13\,°C$ (c 54 °F) and a maximum summer temperature in the shade (because the concrete is shielded by roofing felt and insulation) is about 30 °C (c 86 °F), the maximum increase in temperature is 17 °C. In winter the building may be heated and, even when a temporary heating breakdown occurs, the temperature does not fall below about 5 °C (c 41 °F), so the maximum decrease in temperature is 8 °C. Therefore, the approximate maximum outward movement of the ring beam is

$$\left[0.000\,01 \times \left(\frac{30}{2} + 0.3 \right) \times 17 \right] \times 1000 = 2.6\,\text{mm (0.10 in)}$$

and the approximate maximum inward movement of the ring beam is

$$\left[0.000\,01 \times \left(\frac{30}{2} + 0.3 \right) \times 8 \right] \times 1000 = 1.2\,\text{mm (0.05 in)}$$

Example 3.14 Determine the thermal movement of the dome of Example 3.1 if over a tank.

If in Example 3.1 the concrete is cast at a temperature of about 13 °C (c 54 °F) and a maximum summer temperature in the sun (as the concrete is not shielded) is say 49 °C (c 120 °F), the maximum increase in temperature is 36 °C. In winter the temperature may fall to, say, about -3 °C (c 26 °F), so the maximum decrease in temperature is 16 °C. Therefore the approximate maximum outward movement of the ring beam is

$$\left[0.000\,01 \times \left(\frac{30}{2} + 0.3 \right) \times 36 \right] \times 1000 = 5.5\,\text{mm (0.22 in)}$$

and the approximate maximum inward movement of the ring beam is

$$\left[0.000\,01 \times \left(\frac{30}{2} + 0.3 \right) \times 16 \right] \times 1000 = 2.4\,\text{mm (0.09 in)}$$

57

3.8 Shrinkage Stresses

Shrinkage of concrete is a complicated problem.[3.2] The reinforcement does not shrink, so neither does the structure. The shrinkage stresses are relieved, where high, by multitudes of cracks, many minute (cannot be seen with the naked eye, for example, less than 0.008 mm; 0.000 3 in). Then there is differential shrinkage due to the surface drying out faster than the inner part of the concrete. Sunshine and/or wind on freshly placed concrete can have a most noticeable effect. This often results in surface cracking, which is generally widespread and small in width, but can often be seen with the naked eye.

With domes, as with most structural concrete work, the shrinkage is ignored in the design, since it is considered that relief is provided by cracking, which is tolerated.

References

3.1 Crom J M 1950 Design of prestressed tanks. *Proceedings American Society of Civil Engineers.* Vol 76, Sanitary Engineering Division, New York, ASCE

3.2 Wilby C B 1991 *Concrete materials and structures.* Cambridge–New York, Cambridge University Press

3.3 Timoshenko S and Woinsky-Krieger S 1959 *Theory of plates and shells.* New York, McGraw-Hill

4 Construction

4.1 Protection of Concrete

In the UK, climate and pollution is detrimental to exposed structural concrete. Many materials sometimes used in concrete in the past, but which should not now be used,[4.1] have given exposed concrete bad publicity.

Since about 1945 it was common practice to protect structural concrete in buildings with brickwork, for example, the 11 in (279 mm) cavity-wall used in those days would often be such that its outer 4.5 in (114 mm) skin would protect the structural frame of a building. Then the era of glass curtain walling also protected the structural frame of a building.

The advent of precast concrete buildings, in which the members were exposed to the weather, led to the practice of not always protecting structural concrete from the weather. This led to many problems, resulting from the power of steel corrosion of the reinforcement[4.1] to fracture its surrounding concrete, resulting in further reinforcement corrosion and significant weakening of the structure.

In the UK clients are generally sensitive to the appearance of cracks and rust appearing from the reinforcement.

However, reinforced concrete retaining walls, for example to bridge abutments, are usually exposed to the weather. The same applies to reinforced concrete dams, piles to maritime piers and jetties, and many tanks, containers and water towers. Domes over water and chemical tanks are not always protected against the weather. Domes are sometimes used over sewage tanks to restrain smells and weathertightness does not matter. In these instances the structural concrete of the dome may well be exposed to the weather.

4.2 Construction Details

As an early designer of shells in the UK, the author has generally protected concrete from the weather and used the edge/ring beam to be adequately upstand, as shown by Types B and C, Figs 2.2 and 2.3, to act as a gutter for the rainwater, so this of course has to be watertight. In the case of domes

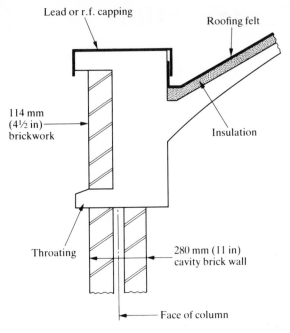

Figure 4.1

this can be done as illustrated in Figs 4.1 and 4.2. Type C is excessive in its provision of a gutter for some rainfall areas and often needs stiffeners against buckling of the ring beam, but it was still used in such rainfall areas where the client preferred not to see any downstand beam.

Abroad there are often vastly different climatic conditions to that in the UK, and some countries have a general practice of exposing concrete to the weather and some apparently tolerate visual cracking. In which case, the shells of Types A, B and C, see Figs 2.1–2.3, might be used fully exposed.

Sometimes domes over tanks are painted. 'Suitable paints[4.2] include rubber base paints, polychloride vinyl-latex, or polymeric vinyl-acrylic paints'. Figures 4.3 and 4.4 show construction details that expose concrete to a modest extent.

Type A in Fig. 2.1 does not incorporate a gutter for rainwater collection. In some locations overseas, where it seldom rains, a gutter is not provided and a walkway or other construction below collects the occasional, rare, rainfall (for example, see Fig. 4.5). Otherwise, some guttering arrangement is sometimes provided at the side of, or just beneath, the ring beam.

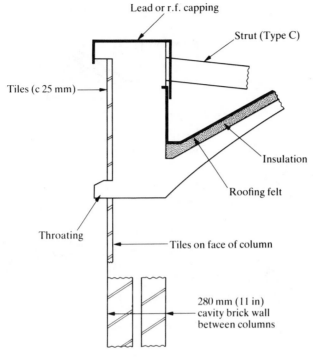

Lead or r.f. capping

Strut (Type C)

Tiles (c 25 mm)

Insulation

Roofing felt

Throating

Tiles on face of column

280 mm (11 in) cavity brick wall between columns

Figure 4.2

4.3 Protection of Prestressing Tendons

Tendons (wires and strands, refer to Section 3.3), which are provided externally, and steel anchorages of tendons are often protected with shotcrete (USA), gunite, pneumatic mortar and modern plastic materials.

Crom[4.3] advocated pneumatic mortar and said: 'Steel wire and rods coated with pneumatic mortar have been free from rust after exposure to tidal water for more than 25 years. There is no case on record of rusting having taken place where steel has been covered with a coating of good pneumatic mortar 0.5 inch (12.7 mm) or more in thickness', and quoted confirmation by tests made at the University of Toronto, Canada.

Shortly after 1950 Curzon Dobell, Vice-President, Preload Enterprises Inc, USA, stated: 'From the construction of many hundreds of tanks over a 20 year period it has been found that a covering of 0.75 inch (19.1 mm) of pneumatic mortar over the wire on the tank wall and dome ring provides adequate protection of the reinforcing. It has been used in tanks in the Andes Mountains, northern parts of Canada and within reach of the spray of the

61

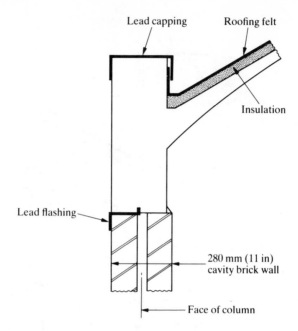

Lead capping

Roofing felt

Insulation

Lead flashing

280 mm (11 in) cavity brick wall

Face of column

Figure 4.3

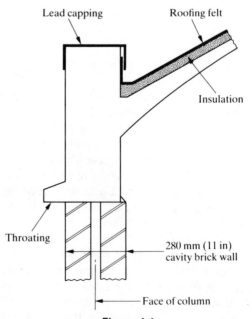

Lead capping

Roofing felt

Insulation

Throating

280 mm (11 in) cavity brick wall

Face of column

Figure 4.4

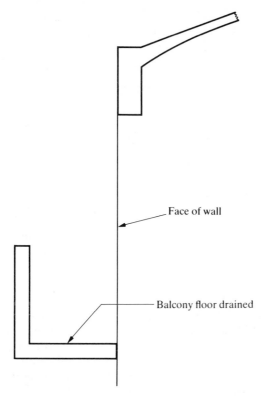

Face of wall

Balcony floor drained

Figure 4.5

Atlantic and Pacific oceans. Where more than one layer of wire is required ... each underlying layer is covered with a thin coat of pneumatic mortar to provide a smooth surface to receive the next layer.'

Claims of a similar nature have been made by a UK manufacturer of ferrocement boats, who used a specially designed concrete with very small aggregate applied to a special steel mesh with a gun.

ACI 344R[4.2] provides 'recommendations not covered by ACI 506, for placing, finishing and curing of shotcrete for walls, roofs and cover coats over prestressing wire', It mentions that the 'total coating thickness over the outside prestressing wire should be not less than 1 in (25 mm)'.

These days modern plastics may sometimes be favoured. The anchorages indicated in Fig. 3.4 were protected with a plastic material.

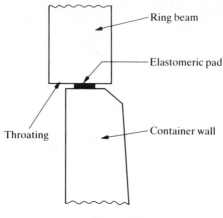

Figure 4.6

4.4 Domes over Tanks

A suitable arrangement is shown in Fig. 4.6. In the section on elastomeric materials, ACI 344R says[4.2], for bearing pads, 'Neoprene or similar suitable materials capable of resistance to prolonged stress and exposure should be used for bearing pads. The neoprene should have a minimum ultimate tensile strength of 2000 psi ($13.8 \, \text{N} \, \text{mm}^{-2}$) and a minimum elongation of 500 per cent (ASTM D412) and a maximum compressive set of 50 per cent (ASTM D395, Method A). Bearing pads are normally 40–60 durometer (ASTM D2240).'

References

4.1 Wilby C B 1991 *Concrete materials and structures.* Cambridge–New York, Cambridge University Press
4.2 ACI Committee 344 Design and construction of circular prestressed concrete structures. *ACI 344R.2R-70.* Detroit, American Concrete Institute 1970 (reaffirmed 1981)
4.3 Crom J M 1950 Design of prestressed tanks. *Proceedings American Society of Civil Engineers.* Vol 76, Sanitary Engineering Division, New York, ASCE

Appendix 1: Wind Pressure on Domes

The most unfavourable wind pressures that could occur on domes in the British Isles, including the Shetland Islands, which are being developed by international oil companies for the North Sea oil business, can be assessed by using the code of practice CP3:Chapter V:Part 2:1972 of the British Standards Institution, London (the figure and table numbers quoted below refer to this code) as follows:

1 The fastest wind speed for the British Isles (from Fig. 1) $= V = 55 \, \text{m s}^{-1}$, (123 miles/hr). (This applies to the Outer Hebrides and the Shetland Islands.)

2 The largest 'topography factor' (Table 2) $= S_1 = 1.1$. (This applies to 'very exposed hill slopes and crests where acceleration of the wind is known to occur, or valleys shaped to produce a funnelling of the wind'.)

3 Values of S_2 relate to 'variation of wind speed with height above ground for various ground roughnesses and building size classes', and are given in Table 3. As a worst case take Class B which is for smaller sizes of buildings and take ground roughness (1) which is for 'Open country with no obstructions.' The height of the roof would perhaps be no higher than, say, 30 m. Then Table 3 gives $S_2 = 1.05$.

4 'The factor S_3 is based on statistical concepts which take account of the degree of security required and the period of time in years during which there will be exposure to wind ... Normally wind loads on completed structures and buildings should be calculated with $S_3 = 1.0$ ...' says the Code, therefore, take $S_3 = 1.0$.

5 The 'design wind speed' $= V_s = VS_1S_2S_3 = 55 \times 1.1 \times 1.05 \times 1.0 = 63.53 \, \text{m s}^{-1}$.

6 The 'dynamic pressure' $= q = kV_s^2$ where $k = 0.613 \, \text{N s}^2 \, \text{m}^{-4}$. Therefore

$$q = 0.613 \times 63.53^2 \, \text{N m}^{-2} = 2.474 \, \text{kN m}^{-2}$$

7 The 'pressure exerted at any point on the surface of a building'

$$p = C_p q = (C_{pe} - C_{pi})q$$

where the external pressure on the roof downwards is $C_{pe}q$ and the internal pressure inside the roof pushing the roof upwards is $C_{pi}q$ so

65

that p is the resultant pressure downwards. Naturally in the following, negative signs for C_{pe} and C_{pi} change pressures into suctions.

8 From Table 8, if, say, the spherical dome has edges (springing points) at 40° on either side of the vertical through its crown, then, taking the more unfavourable very high compared to width buildings

    (a) on the windward side:

        (i) at the springing point, roof surface inclination to horizontal = 40°, $C_{pe} = -0.2$

        (ii) at the crown, roof surface inclination to horizontal = 0°, $C_{pe} = -0.7$

    (b) on the leeward side:

        (i) at the springing point, roof surface inclination to horizontal = 40°, $C_{pe} = -0.5$

        (ii) at the crown, roof surface inclination to horizontal = 0°, $C_{pe} = -0.6$.

9 From Appendix E of the code, the largest positive value of $C_{pi} = 0.6$, when a dominant opening on the windward side makes the permeability of this face greater than three times the total distributed permeability of all the faces subject to suction.

'The distributed permeability should be assessed in each case as accurately as is practicable. As a guide it can be said that the typical permeability of a house or office block with all windows nominally closed is in the range of 0.01–0.05 per cent of the face area, depending on the degree of draughtproofing.'

10 From Appendix E of the code, the largest negative value of $C_{pi} = -0.9$, when a dominant opening on a face parallel to the wind has an area greater than three times the total distributed permeability of the total other distributed permeability of all the external faces subject to suction.

11 The greatest possible uplift pressure is given by using $C_{pe} = -0.7$ from [8(a)(ii)], $C_{pi} = +0.6$ from (9) and $q = 2.474\,\text{kN m}^{-2}$ from (6) in the equation given in (7) previously

$$p = (-0.7 - 0.6) \times 2.474 = -3.22\,\text{kN m}^{-2}$$

where the negative sign indicates suction on the top of the dome, i.e. negative resultant pressure downwards.

12 The greatest possible downward pressure is given by using $C_{pe} = -0.2$ from [8(a)(i)], $C_{pi} = -0.9$ from (10) and $q = 2.666\,\text{kN m}^{-2}$ from (6) in the equation given in (7) previously

$$p = (-0.2 + 0.9) \times 2.474 = +1.73\,\text{kN m}^{-2}$$

where the positive sign indicates pressure on the top of the dome, i.e. positive resultant pressure downwards.

13 With most buildings (9) and (10) previously would not apply so (11)

previously would give

$$p = -0.7 \times 2.474 = -1.73 \, \text{kN m}^{-2}$$

i.e. $1.73 \, \text{kN m}^{-2}$ as the greatest possible resultant uplift pressure, and (12) previously would give

$$p = -0.2 \times 2.474 = -0.495 \, \text{kN m}^{-2}$$

i.e. there is never any resultant downward pressure.

One cannot really conceive of any building or tank, etc. where domes are used, where any one of the structures would have a large dominant opening on either the windward or leeward side. Thus (11) and (12) above would only apply to a dome say peculiarly used as an aircraft hangar, that is with a very large opening at one side—the author does not know of any dome used in this way. Therefore, this present section is relevant in practice instead of (11) and (12) previously.

Internationally, the Beaufort Scale defines a 'storm' as comprising wind velocities of $28.7–33.5 \, \text{m s}^{-1}$ (64–75 miles/hr) and a 'hurricane' as having wind velocities greater than $33.5 \, \text{m s}^{-1}$ (75 miles/hr), so the above example which considers a basic wind velocity of $55 \, \text{m s}^{-1}$ (123 miles/hr) and a design wind velocity of $63.53 \, \text{m s}^{-1}$ (142 miles/hr) should be more severe than for most places in the world.

For England and Wales the fastest wind speed (from Fig. 1) $= V = 48 \, \text{m s}^{-1}$ (108 miles/hr). Taking unfavourable conditions as previously $S_1 = 1.1$, $S_2 = 1.05$ and $S_3 = 1.0$. Then the 'design wind speed' is

$$V_s = V S_1 S_2 S_3 = 48 \times 1.1 \times 1.05 \times 1.0$$

$$= 55.44 \, \text{m s}^{-1} \, (124 \, \text{miles/hr})$$

The 'dynamic pressure' $= q = k V_s^2$ where $k = 0.613 \, \text{N s}^2 \, \text{m}^{-4}$. Therefore

$$q = 0.613 \times 55.44^2 \, \text{N m}^{-2}$$

$$= 1.884 \, \text{kN m}^{-2}$$

Sections (7) and (8) previously are relevant. Then similarly to (13) previously

$$p = -0.7 \times 1.884$$

$$= -1.319 \, \text{kN m}^{-2}$$

i.e. $1.319 \, \text{kN m}^{-2}$ is the greatest possible resultant uplift pressure and

$$p = -0.2 \times 1.884$$

$$= -0.377 \, \text{kN m}^{-2}$$

i.e. there is never any resultant downward pressure.

Appendix 2: Snow Loading on Domes

In the UK the snow load (which takes care of access for cleaning and repair) is specified by the code of practice CP 3:Chapter V:Part 1:1972 of the British Standards Institution, London:

(a) for portions inclined at less than 30° to the horizontal as $0.75 \, \mathrm{kN \, m^{-2}}$ of plan area,
(b) for portions inclined at more than 75° to the horizontal as zero,
(c) for portions inclined between 30° and 75° to the horizontal as linearly interpolated between (a) and (b). For example, for a portion inclined at $\theta°$ to the horizontal the snow load of plan area is

$$0.75 \times \frac{\theta - 30}{75 - 30} = \frac{\theta - 30}{60} \mathrm{kN \, m^{-2}}$$

In effect $0.75 \, \mathrm{kN \, m^{-2}}$ of plan area is satisfactory for all domes in the UK. It is reasonable practice to use $0.75 \, \mathrm{kN \, m^{-2}}$ of curved surface area so that the loading case (i.e. distribution) is the same as that for the self-weight of the dome. This simplifies calculations and is an assumption giving the design slightly greater safety.

For North America, roofs experiencing the most severe snow loads are in Canada. The author is indebted to Professor John Christian of the Memorial University of Newfoundland, St John's, Canada, for the information which follows on snow loads on roofs in Canada. Just as the British code of practice for wind loading incorporates a map specifying the wind speeds in different locations of the British Isles, the Canadian code (National Building Code of Canada, 1975) incorporates a map specifying snow loads on the ground as basic pressures to which the design of roofs should be related in different locations of Canada. Extracts from this map are approximately as follows:

(a) Calgary $20 \, \mathrm{lb/ft^2}$ $(1 \, \mathrm{kN \, m^{-2}})$ increasing to $120 \, \mathrm{lb/ft^2}$ $(6 \, \mathrm{kN \, m^{-2}})$ towards the Rocky Mountains,
(b) Ottawa $60 \, \mathrm{lb/ft^2}$ $(3 \, \mathrm{kN \, m^{-2}})$,
(c) Quebec $84 \, \mathrm{lb/ft^2}$ $(4.2 \, \mathrm{kN \, m^{-2}})$,
(d) St John's, Newfoundland, $30\text{–}40 \, \mathrm{lb/ft^2}$ $(1.5\text{–}2 \, \mathrm{kN \, m^{-2}})$ increasing to $120 \, \mathrm{lb/ft^2}$ $(6 \, \mathrm{kN \, m^{-2}})$ towards North Labrador,

(e)   Toronto 40 lb/ft^2 (2 kN m^{-2}),

(f)   Montreal 54 lb/ft^2 (2.7 kN m^{-2}),

(g)   northern and eastern parts of Canada generally up to about 100 lb/ft^2 (5 kN m^{-2}),

(h)   generally less towards the USA, but yet as great as 80 lb/ft^2 (4 kN m^{-2}) at the border with the USA in the Rocky Mountains.

In the USA there is, for example, 80 lb/ft^2 (4 kN m^{-2}) in the Rocky Mountains next to Canada, but yet parts in the south that never experience snow.

As regards the snow loads for which roofs should be designed, these would normally be 0.8 of the above values of 'ground snow loads' specified for North America, according to the National Building Code of Canada.

The *Building Code Requirements for Minimum Design Loads in Buildings and Other Structures.* USASI A-59.1, USA Standards Institute, New York, 1971, incorporates a map of the USA giving the 'estimated weight of seasonal snowpack equalled or exceeded one year in ten'. This map excludes zones which seemingly include the Rocky Mountains. The legend to the map gives zones with less than 10 psf (lb/ft^2) of plan area and with 10, 20, 30 and 40 psf respectively. These are snow.loads to be used (except that more detailed information on snow loads is usually available locally) in the design of roofs except that a minimum roof load of 20 psf is specified to provide for construction and repair loads and to ensure reasonable stiffness. The above-mentioned map for the USA is also given in Winter and Nilson 1972 *Design of concrete structures.* McGraw-Hill, New York.

Appendix 3: Conversion Table

For the purpose of absolute clarity of international units used in this book, the following conversions (which should prove useful to engineers) are given.

British Imperial	U.S.A.	Metric	SI
1 ton	1 long ton	1016.0 kg	9.964 kN
2000 lb	1 short ton	907.1 kg	8.896 kN
0.9843 ton	0.9843 long tons	1 tonne 1000 kg	9.807 kN
1 lb	1 lb	0.4536 kg	4.448 N
1000 lb	1 kip	453.6 kg	4.448 kN
1 inch	1 inch	2.54 cm	25.4 mm
1 foot	1 foot	30.48 cm	0.3048 m
1000 lb/in	1 kip/in	178.6 kg cm^{-1}	175.1 kN m^{-1}
1 lb/in^2	1 psi	0.070 309 kg cm^{-2}	6.895 kN m^{-2}
1000 lb/in^2	1 kip/in^2 1000 psi	70.309 kg cm^{-2}	6.895 N mm^{-2}
1 lb/ft^2	1 lb/ft^2	4.882 kg m^{-2}	0.047 88 kN m^{-2}
1 ton/ft^2	1 long ton/ft^2	10940 kg m^{-2}	107.3 kN m^{-2}
1 lb/ft	1 lb/ft	1.488 kg m^{-1}	0.014 59 kN m^{-1}
1 ton/ft	1 long ton/ft	3333 kg m^{-1}	32.69 kN m^{-1}
1 lb/ft^3	1 lb/ft^3	16.02 kg m^{-3}	0.157 07 kN m^{-3}
	1 Pa	10.20 kg cm^{-2}	1 N m^{-2}

Notes
1. The terms 'force' and 'mass' have not been used above, and acceleration due to gravity = 9.807 m s^{-2}
2. p.s.i. = psi = lb/in^2 = pounds per square inch
3. p.s.f. = psf = lb/ft^2 = pounds per square foot
4. kip = 1000 lb = 1000 pounds
5. kip/in^2 = 1000 psi = 1000 pounds per square inch
6. Pa = pascal

Appendix 4: Nomenclature for Non-prestressed and Prestressed Dome Design Tables (Appendices 5 and 6)

The symbols shown on the tables are as follows:

A   curved surface area of soffit of the shell. This is useful for estimating the cost of the shuttering (formwork)

ALPHA   angle α, see Fig. 2.4

b   breadth of ring beam, see Fig. 2.5

d   depth of ring beam, see Fig. 2.5

db   see Fig. 2.5

DIA.   shell diameter or span $= 2r$ in Fig. 1.2, see Fig. 2.4

$fb = f_b$ bottom fibre stress in ring beam, positive sign indicates tension

$ft = f_t$ top fibre stress in ring beam, positive sign indicates tension

h1   thickness of shell away from thickening towards supports, see Fig. 2.5

h2   thickness of thickening of shell at springing line, see Fig. 2.5

$Mf = M_\phi$ positive sign indicates tension on bottom face of shell

$Nf = N_\phi$ positive sign indicates tension

$Nt = N_\theta$ positive sign indicates tension

p   curved length of soffit of shell from springing line to crown to springing line at diametrically opposite side, see Fig. 2.4

qb   safe buckling load in $\mathrm{kN\,m}^{-2}$ of curved surface area of the shell, see Section 2.3

qd   self-weight of shell plus 30 per cent of the weight of its thickening spread over the curved surface of the shell, in $\mathrm{kN\,m}^{-2}$ of curved surface area, see Section 2.4

qD   self-weight of shell plus all of the weight of its thickening spread over the curved surface of the shell, in $\mathrm{kN\,m}^{-2}$ of curved surface area, see Section 2.4

$Qf = N_\phi$ positive sign indicates directions shown in Fig. 2.6

RADIUS   radius of soffit of shell part of dome, see Fig. 2.4

RISE   height of soffit of crown above springing line of shell part of
          dome, see Fig. 2.4
s   curved distance from springing line, see Fig. 2.5
S   curved length of thickening, see Fig. 2.5
Vol.   volume of the shell plus its thickenings. This is useful for
          estimating the cost of the concrete for the shell part of the dome
y   see Fig. 2.5

Appendix 5: Design Tables for Non-prestressed Domes

Tables 1A to 40A For Type A Domes

```
TABLE  1  A
   DIA.          RISE          RADIUS         ALPHA
    m             m              m            deg.
 12.000         1.496         12.779         28.00
   h1      h2         S            b             d
   mm      mm         m            mm            mm
   75      180      2.553         250           700
    y      db        qd            qb            p
   mm      mm     kN/sq.m       kN/sq.m          m
  189     496      2.03         13.78         12.491
    A      qD         ft            fb           Vol.
  sq.m   kN/sq.m   N/sq.mm       N/sq.mm        cu.m
  120     2.636     0.076         0.166         13.4
    s      Nt         Mf            Nf            Qf
    m      kN/m     kNm/m         kN/m          kN/m
  0.00    13.7     -0.42         -4.61         -1.16
  0.45     8.7     -0.03         -5.51         -0.59
  0.90     3.3      0.14         -6.20         -0.21
  1.35    -1.1      0.19         -6.63          0.00
  1.80    -4.1      0.16         -6.84          0.09
  2.25    -5.7      0.11         -6.89          0.11
  2.70    -6.5      0.07         -6.84          0.09
  3.60    -6.7      0.01         -6.63          0.04
  4.50    -6.5     -0.01         -6.45          0.00
```

TABLE 2 A

DIA.		RISE		R@DIUS		ALPHA
m		m		m		deg.
12.400		1.546		13.205		28.00

h1	h2	S	b	d
mm	mm	m	mm	mm
75	180	2.595	250	700

y	db	qd	qb	p
mm	mm	kN/sq.m	kN/sq.m	m
189	496	2.03	12.90	12.908

A	qD	ft	fb	Vol.
sq.m	kN/sq.m	N/sq.mm	N/sq.mm	cu.m
128	2.624	0.081	0.177	14.3

s	Nt	Mf	Nf	Qf
m	kN/m	kNm/m	kN/m	kN/m
0.00	14.7	-0.43	-4.76	-1.20
0.45	9.4	-0.03	-5.69	-0.62
0.90	3.7	0.15	-6.40	-0.23
1.35	-0.9	0.20	-6.85	0.00
1.80	-4.0	0.17	-7.07	0.09
2.25	-5.9	0.12	-7.13	0.12
2.70	-6.7	0.07	-7.08	0.10
3.15	-7.0	0.04	-6.98	0.07
3.60	-7.0	0.01	-6.87	0.04
4.50	-6.7	-0.01	-6.67	0.01

TABLE 3 A

DIA.	RISE	RADIUS	ALPHA
m	m	m	deg.
12.700	1.583	13.524	28.00

h1	h2	S	b	d
mm	mm	m	mm	mm
75	180	2.626	250	700

y	db	qd	qb	p
mm	mm	kN/sq.m	kN/sq.m	m
189	496	2.02	12.30	13.220

A	qD	ft	fb	Vol.
sq.m	kN/sq.m	N/sq.mm	N/sq.mm	cu.m
135	2.615	0.085	0.186	14.9

s	Nt	Mf	Nf	Qf
m	kN/m	kNm/m	kN/m	kN/m
0.00	15.4	-0.44	-4.88	-1.23
0.45	9.9	-0.03	-5.82	-0.64
0.90	4.0	0.16	-6.55	-0.24
1.35	-0.7	0.21	-7.01	-0.01
1.80	-4.0	0.18	-7.24	0.10
2.25	-5.9	0.13	-7.30	0.12
2.70	-6.9	0.08	-7.26	0.10
3.15	-7.2	0.04	-7.16	0.07
4.05	-7.0	0.00	-6.94	0.02
4.95	-6.8	-0.01	-6.78	0.00

TABLE 4 A

DIA.	RISE	RADIUS	ALPHA
m	m	m	deg.
13.100	1.633	13.950	28.00

h1	h2	S	b	d
mm	mm	m	mm	mm
75	180	2.667	250	700

y	db	qd	qb	p
mm	mm	kN/sq.m	kN/sq.m	m
189	496	2.02	11.56	13.636

A	qD	ft	fb	Vol.
sq.m	kN/sq.m	N/sq.mm	N/sq.mm	cu.m
143	2.604	0.090	0.197	15.8

s	Nt	Mf	Nf	Qf
m	kN/m	kNm/m	kN/m	kN/m
0.00	16.5	-0.46	-5.03	-1.27
0.45	10.7	-0.03	-6.00	-0.66
0.90	4.5	0.17	-6.75	-0.25
1.35	-0.5	0.22	-7.23	-0.01
1.80	-4.0	0.20	-7.47	0.10
2.25	-6.0	0.14	-7.54	0.13
2.70	-7.0	0.09	-7.50	0.11
3.15	-7.4	0.04	-7.40	0.08
4.05	-7.3	0.00	-7.17	0.02
4.95	-7.0	-0.01	-7.00	0.00

TABLE 5 A

DIA.	RISE	RADIUS	ALPHA
m	m	m	deg.
13.500	1.683	14.376	28.00

h1	h2	S	b	d
mm	mm	m	mm	mm
75	180	2.708	250	700

y	db	qd	qb	p
mm	mm	kN/sq.m	kN/sq.m	m
189	496	2.02	10.89	14.053

A	qD	ft	fb	Vol.
sq.m	kN/sq.m	N/sq.mm	N/sq.mm	cu.m
152	2.594	0.096	0.209	16.7

s	Nt	Mf	Nf	Qf
m	kN/m	kNm/m	kN/m	kN/m
0.00	17.5	-0.47	-5.18	-1.31
0.45	11.4	-0.03	-6.18	-0.69
0.90	5.0	0.18	-6.95	-0.26
1.35	-0.3	0.23	-7.44	-0.01
1.80	-3.9	0.21	-7.70	0.10
2.25	-6.1	0.15	-7.77	0.13
2.70	-7.2	0.10	-7.74	0.12
3.15	-7.6	0.05	-7.64	0.09
4.05	-7.5	0.00	-7.40	0.03
4.95	-7.2	-0.01	-7.23	0.00

```
TABLE  6  A
   DIA.           RISE          RADIUS         ALPHA
     m              m              m            deg.
  13.900          1.733         14.802         28.00
    h1       h2        S          b             d
    mm       mm        m          mm            mm
    75       180       2.748      250           700
    y        db        qd         qb            p
    mm       mm     kN/sq.m    kN/sq.m          m
   189       496     2.01       10.27         14.469
    A        qD        ft         fb           Vol.
  sq.m    kN/sq.m   N/sq.mm    N/sq.mm         cu.m
   161     2.583     0.102      0.222          17.6
    s        Nt        Mf         Nf            Qf
    m       kN/m     kNm/m       kN/m          kN/m
  0.00     18.6     -0.48       -5.33         -1.35
  0.45     12.2     -0.03       -6.36         -0.71
  0.90      5.5      0.19       -7.15         -0.27
  1.35      0.0      0.25       -7.66         -0.02
  1.80     -3.9      0.22       -7.93          0.10
  2.25     -6.2      0.17       -8.01          0.14
  2.70     -7.4      0.11       -7.97          0.13
  3.15     -7.8      0.06       -7.88          0.09
  3.60     -7.9      0.02       -7.75          0.06
  4.50     -7.6     -0.01       -7.53          0.01
  5.40     -7.4     -0.01       -7.39          0.00
```

TABLE 7 A

DIA.	RISE	RADIUS	ALPHA
m	m	m	deg.
14.300	1.783	15.228	28.00

h1	h2	S	b	d
mm	mm	m	mm	mm
75	180	2.787	250	700

y	db	qd	qb	p
mm	mm	kN/sq.m	kN/sq.m	m
189	496	2.01	9.70	14.886

A	qD	ft	fb	Vol.
sq.m	kN/sq.m	N/sq.mm	N/sq.mm	cu.m
171	2.573	0.108	0.234	18.6

s	Nt	Mf	Nf	Qf
m	kN/m	kNm/m	kN/m	kN/m
0.00	19.7	-0.50	-5.48	-1.39
0.45	13.0	-0.02	-6.54	-0.74
0.90	6.0	0.20	-7.35	-0.29
1.35	0.3	0.26	-7.88	-0.02
1.80	-3.8	0.24	-8.15	0.11
2.25	-6.3	0.18	-8.25	0.14
2.70	-7.6	0.12	-8.21	0.13
3.15	-8.1	0.06	-8.12	0.10
3.60	-8.1	0.03	-7.99	0.07
4.50	-7.8	-0.01	-7.76	0.02
5.40	-7.6	-0.01	-7.62	0.00

TABLE 8 A

DIA.		RISE		RADIUS		ALPHA
m		m		m		deg.
14.800		1.845		15.761		28.00
h1	h2		S		b	d
mm	mm		m		mm	mm
75	180		2.835		250	700
y	db		qd		qb	p
mm	mm		kN/sq.m		kN/sq.m	m
189	496		2.01		9.06	15.406
A	qD		ft		fb	Vol.
sq.m	kN/sq.m		N/sq.mm		N/sq.mm	cu.m
183	2.561		0.116		0.250	19.8
s	Nt		Mf		Nf	Qf
m	kN/m		kNm/m		kN/m	kN/m
0.00	21.2		-0.51		-5.67	-1.44
0.45	14.1		-0.02		-6.76	-0.77
0.90	6.7		0.21		-7.60	-0.30
1.35	0.7		0.28		-8.15	-0.03
1.80	-3.7		0.26		-8.44	0.11
2.25	-6.3		0.20		-8.54	0.15
2.70	-7.8		0.13		-8.51	0.14
3.15	-8.3		0.07		-8.42	0.11
3.60	-8.4		0.03		-8.29	0.07
4.50	-8.1		-0.01		-8.05	0.02
5.40	-7.8		-0.01		-7.90	0.00

TABLE 9 A

DIA.	RISE	RADIUS	ALPHA
m	m	m	deg.
15.200	1.895	16.186	28.00

h1	h2	S	b	d
mm	mm	m	mm	mm
75	180	2.873	250	700

y	db	qd	qb	p
mm	mm	kN/sq.m	kN/sq.m	m
189	496	2.00	8.59	15.822

A	qD	ft	fb	Vol.
sq.m	kN/sq.m	N/sq.mm	N/sq.mm	cu.m
193	2.552	0.122	0.264	20.8

s	Nt	Mf	Nf	Qf
m	kN/m	kNm/m	kN/m	kN/m
0.00	22.3	-0.52	-5.82	-1.48
0.45	14.9	-0.02	-6.94	-0.80
0.90	7.3	0.22	-7.80	-0.32
1.35	1.0	0.30	-8.36	-0.03
1.80	-3.6	0.27	-8.67	0.11
2.25	-6.4	0.21	-8.78	0.16
2.70	-7.9	0.14	-8.75	0.15
3.15	-8.6	0.08	-8.66	0.12
3.60	-8.7	0.04	-8.53	0.08
4.50	-8.4	-0.01	-8.28	0.02
5.85	-8.0	-0.01	-8.08	-0.01

TABLE 10 A

DIA.	RISE	RADIUS	ALPHA
m	m	m	deg.
15.700	1.957	16.719	28.00

h1	h2	S	b	d
mm	mm	m	mm	mm
75	180	2.920	250	700

y	db	qd	qb	p
mm	mm	kN/sq.m	kN/sq.m	m
189	496	2.00	8.05	16.343

A	qD	ft	fb	Vol.
sq.m	kN/sq.m	N/sq.mm	N/sq.mm	cu.m
206	2.541	0.130	0.281	22.1

s	Nt	Mf	Nf	Qf
m	kN/m	kNm/m	kN/m	kN/m
0.00	23.9	-0.54	-6.00	-1.53
0.45	16.1	-0.02	-7.16	-0.83
0.90	8.1	0.24	-8.04	-0.34
1.35	1.4	0.32	-8.63	-0.04
1.80	-3.4	0.29	-8.95	0.11
2.25	-6.4	0.23	-9.07	0.16
2.70	-8.1	0.15	-9.05	0.16
3.15	-8.8	0.09	-8.96	0.13
3.60	-9.0	0.04	-8.83	0.09
4.05	-8.9	0.01	-8.69	0.05
4.95	-8.5	-0.01	-8.48	0.01
5.85	-8.3	-0.01	-8.36	-0.01

TABLE 11 A

DIA.		RISE		RADIUS		ALPHA
m		m		m		deg.
16.100		2.007		17.145		28.00

h1	h2	S	b	d
mm	mm	m	mm	mm
75	180	2.957	250	700

y	db	qd	qb	p
mm	mm	kN/sq.m	kN/sq.m	m
189	496	2.00	7.65	16.759

A	qD	ft	fb	Vol.
sq.m	kN/sq.m	N/sq.mm	N/sq.mm	cu.m
216	2.533	0.137	0.295	23.2

s	Nt	Mf	Nf	Qf
m	kN/m	kNm/m	kN/m	kN/m
0.00	25.1	-0.55	-6.15	-1.57
0.45	17.0	-0.01	-7.33	-0.85
0.90	8.7	0.25	-8.24	-0.35
1.35	1.8	0.33	-8.85	-0.04
1.80	-3.3	0.31	-9.18	0.11
2.25	-6.5	0.24	-9.31	0.17
2.70	-8.3	0.17	-9.29	0.17
3.15	-9.1	0.10	-9.20	0.13
3.60	-9.3	0.05	-9.07	0.09
4.05	-9.1	0.01	-8.93	0.06
4.95	-8.7	-0.01	-8.71	0.01
5.85	-8.5	-0.01	-8.58	-0.01

```
TABLE  12  A
```

DIA.	RISE	RADIUS	ALPHA
m	m	m	deg.
16.600	2.070	17.677	28.00

h1	h2	S	b	d
mm	mm	m	mm	mm
75	180	3.003	250	700

y	db	qd	qb	p
mm	mm	kN/sq.m	kN/sq.m	m
189	496	2.00	7.20	17.280

A	qD	ft	fb	Vol.
sq.m	kN/sq.m	N/sq.mm	N/sq.mm	cu.m
230	2.523	0.146	0.312	24.6

s	Nt	Mf	Nf	Qf
m	kN/m	kNm/m	kN/m	kN/m
0.00	26.7	-0.57	-6.33	-1.62
0.45	18.2	-0.01	-7.55	-0.89
0.90	9.5	0.27	-8.49	-0.37
1.35	2.2	0.35	-9.11	-0.05
1.80	-3.1	0.33	-9.47	0.12
2.25	-6.5	0.26	-9.60	0.18
2.70	-8.4	0.18	-9.59	0.18
3.15	-9.3	0.11	-9.50	0.14
3.60	-9.6	0.06	-9.37	0.10
4.05	-9.5	0.02	-9.23	0.06
4.95	-9.0	-0.01	-9.00	0.01
6.30	-8.7	-0.01	-8.82	-0.01

TABLE 13 A

DIA. m		RISE m		RADIUS m		ALPHA deg.
17.100		2.132		18.210		28.00

h1 mm	h2 mm	S m		b mm		d mm
75	180	3.047		250		700

y mm	db mm	qd kN/sq.m		qb kN/sq.m		p m
189	496	1.99		6.79		17.800

A sq.m	qD kN/sq.m	ft N/sq.mm	fb N/sq.mm	Vol. cu.m
244	2.513	0.155	0.331	26.0

s m	Nt kN/m	Mf kNm/m	Nf kN/m	Qf kN/m
0.00	28.4	-0.58	-6.52	-1.68
0.45	19.5	0.00	-7.77	-0.92
0.90	10.4	0.28	-8.74	-0.39
1.35	2.7	0.38	-9.38	-0.05
1.80	-2.9	0.36	-9.75	0.12
2.25	-6.5	0.28	-9.90	0.18
2.70	-8.6	0.20	-9.90	0.18
3.15	-9.6	0.12	-9.80	0.15
3.60	-9.9	0.06	-9.67	0.11
4.05	-9.8	0.02	-9.53	0.07
5.40	-9.1	-0.02	-9.20	0.00
6.30	-8.9	-0.01	-9.10	-0.01

TABLE 14 A

DIA. m	RISE m	RADIUS m	ALPHA deg.
17.600	2.194	18.742	28.00

h1 mm	h2 mm	S m	b mm	d mm
75	180	3.092	250	700

y mm	db mm	qd kN/sq.m	qb kN/sq.m	p m
189	496	1.99	6.41	18.321

A sq.m	qD kN/sq.m	ft N/sq.mm	fb N/sq.mm	Vol. cu.m
258	2.504	0.164	0.349	27.4

s m	Nt kN/m	Mf kNm/m	Nf kN/m	Qf kN/m
0.00	30.1	-0.59	-6.70	-1.73
0.45	20.7	0.00	-7.99	-0.95
0.90	11.2	0.30	-8.98	-0.41
1.35	3.3	0.40	-9.65	-0.06
1.80	-2.6	0.38	-10.03	0.12
2.25	-6.5	0.31	-10.19	0.19
2.70	-8.8	0.22	-10.20	0.19
3.15	-9.9	0.14	-10.11	0.16
3.60	-10.2	0.07	-9.97	0.12
4.05	-10.1	0.03	-9.82	0.08
4.50	-9.9	0.00	-9.69	0.04
5.40	-9.4	-0.02	-9.48	0.00
6.75	-9.2	-0.01	-9.34	-0.01

TABLE 15 A

DIA.	RISE	RADIUS	ALPHA
m	m	m	deg.
18.000	2.244	19.168	28.00

h1	h2	S	b	d
mm	mm	m	mm	mm
90	200	3.334	280	800

y	db	qd	qb	p
mm	mm	kN/sq.m	kN/sq.m	m
224	573	2.37	8.82	18.737

A	qD	ft	fb	Vol.
sq.m	kN/sq.m	N/sq.mm	N/sq.mm	cu.m
270	2.929	0.136	0.283	33.6

s	Nt	Mf	Nf	Qf
m	kN/m	kNm/m	kN/m	kN/m
0.00	27.3	-0.75	-6.85	-1.77
0.45	19.8	-0.12	-8.02	-1.05
0.90	11.6	0.23	-8.97	-0.52
1.35	4.4	0.38	-9.65	-0.16
1.80	-1.3	0.40	-10.09	0.05
2.25	-5.3	0.35	-10.32	0.16
2.70	-7.9	0.27	-10.39	0.19
3.15	-9.4	0.19	-10.36	0.17
3.60	-10.1	0.12	-10.26	0.14
4.05	-10.3	0.06	-10.13	0.10
4.50	-10.2	0.02	-10.00	0.07
5.85	-9.6	-0.02	-9.69	0.00
6.75	-9.4	-0.01	-9.58	-0.01

```
TABLE  16 A
 DIA.           RISE          RADIUS          ALPHA
  m              m               m             deg.
18.500         2.307         19.701          28.00
  h1        h2        S           b              d
  mm        mm        m           mm             mm
  90        200      3.380        280            800
  y         db        qd          qb             p
  mm        mm     kN/sq.m     kN/sq.m           m
 224        573      2.36        8.35         19.258
  A         qD        ft          fb            Vol.
sq.m     kN/sq.m   N/sq.mm     N/sq.mm          cu.m
 286       2.920    0.144       0.299           35.3
  s         Nt        Mf          Nf             Qf
  m        kN/m     kNm/m       kN/m            kN/m
0.00       28.9     -0.76       -7.03          -1.82
0.45       21.0     -0.11       -8.24          -1.09
0.90       12.5      0.25       -9.21          -0.54
1.35        5.0      0.40       -9.92          -0.17
1.80       -1.0      0.42      -10.37           0.05
2.25       -5.2      0.37      -10.61           0.16
2.70       -8.0      0.29      -10.69           0.19
3.15       -9.6      0.20      -10.66           0.18
3.60      -10.4      0.13      -10.56           0.15
4.05      -10.6      0.07      -10.43           0.11
4.50      -10.5      0.03      -10.30           0.07
5.85       -9.9     -0.02       -9.97           0.01
7.20       -9.7     -0.01       -9.82          -0.01
```

TABLE 17 A

DIA.	RISE	RADIUS	ALPHA
m	m	m	deg.
19.100	2.381	20.340	28.00

h1	h2	S	b	d
mm	mm	m	mm	mm
90	200	3.435	280	800

y	db	qd	qb	p
mm	mm	kN/sq.m	kN/sq.m	m
224	573	2.36	7.83	19.882

A	qD	ft	fb	Vol.
sq.m	kN/sq.m	N/sq.mm	N/sq.mm	cu.m
304	2.909	0.153	0.318	37.5

s	Nt	Mf	Nf	Qf
m	kN/m	kNm/m	kN/m	kN/m
0.00	30.8	-0.79	-7.25	-1.89
0.45	22.5	-0.11	-8.50	-1.13
0.90	13.6	0.26	-9.50	-0.57
1.35	5.7	0.42	-10.23	-0.18
1.80	-0.6	0.45	-10.70	0.05
2.25	-5.1	0.40	-10.95	0.16
2.70	-8.1	0.31	-11.04	0.20
3.15	-9.8	0.22	-11.02	0.19
3.60	-10.7	0.14	-10.92	0.16
4.05	-11.0	0.08	-10.79	0.12
4.50	-10.9	0.03	-10.65	0.08
4.95	-10.7	0.01	-10.52	0.05
5.85	-10.3	-0.02	-10.31	0.01
7.20	-10.0	-0.01	-10.15	-0.01

TABLE 18 A

DIA.	RISE	RADIUS	ALPHA
m	m	m	deg.
19.700	2.456	20.979	28.00

h1	h2	S	b	d
mm	mm	m	mm	mm
90	200	3.488	280	800

y	db	qd	qb	p
mm	mm	kN/sq.m	kN/sq.m	m
224	573	2.36	7.36	20.507

A	qD	ft	fb	Vol.
sq.m	kN/sq.m	N/sq.mm	N/sq.mm	cu.m
324	2.898	0.163	0.337	39.8

s	Nt	Mf	Nf	Qf
m	kN/m	kNm/m	kN/m	kN/m
0.00	32.8	-0.81	-7.47	-1.95
0.45	24.0	-0.11	-8.75	-1.18
0.90	14.8	0.28	-9.79	-0.59
1.35	6.5	0.45	-10.54	-0.20
1.80	-0.2	0.48	-11.03	0.04
2.25	-5.0	0.43	-11.30	0.17
2.70	-8.2	0.34	-11.40	0.21
3.15	-10.1	0.25	-11.38	0.21
3.60	-11.0	0.16	-11.28	0.17
4.05	-11.3	0.09	-11.15	0.13
4.50	-11.3	0.04	-11.01	0.09
4.95	-11.1	0.01	-10.87	0.06
6.30	-10.4	-0.02	-10.58	0.00
7.65	-10.3	-0.01	-10.46	-0.01

TABLE 19 A

DIA.	RISE	RADIUS	ALPHA
m	m	m	deg.
20.300	2.531	21.617	28.00

h1	h2	S	b	d
mm	mm	m	mm	mm
90	200	3.541	280	800

y	db	qd	qb	p
mm	mm	kN/sq.m	kN/sq.m	m
224	573	2.35	6.93	21.131

A	qD	ft	fb	Vol.
sq.m	kN/sq.m	N/sq.mm	N/sq.mm	cu.m
344	2.888	0.173	0.357	42.1

s	Nt	Mf	Nf	Qf
m	kN/m	kNm/m	kN/m	kN/m
0.00	34.9	-0.83	-7.69	-2.01
0.45	25.7	-0.11	-9.01	-1.22
0.90	15.9	0.30	-10.07	-0.62
1.35	7.3	0.48	-10.85	-0.21
1.80	0.3	0.51	-11.36	0.04
2.25	-4.8	0.46	-11.65	0.17
2.70	-8.2	0.37	-11.75	0.22
3.15	-10.3	0.27	-11.74	0.22
3.60	-11.3	0.18	-11.65	0.18
4.05	-11.7	0.10	-11.51	0.14
4.50	-11.7	0.05	-11.37	0.10
4.95	-11.5	0.01	-11.23	0.06
6.30	-10.8	-0.02	-10.92	0.00
7.65	-10.5	-0.01	-10.79	-0.01

TABLE 20 A

DIA.	RISE	RADIUS	ALPHA
m	m	m	deg.
20.900	2.606	22.256	28.00

h1	h2	S	b	d
mm	mm	m	mm	mm
90	200	3.593	280	800

y	db	qd	qb	p
mm	mm	kN/sq.m	kN/sq.m	m
224	573	2.35	6.54	21.756

A	qD	ft	fb	Vol.
sq.m	kN/sq.m	N/sq.mm	N/sq.mm	cu.m
364	2.878	0.183	0.378	44.4

s	Nt	Mf	Nf	Qf
m	kN/m	kNm/m	kN/m	kN/m
0.00	37.0	-0.85	-7.91	-2.08
0.45	27.3	-0.11	-9.27	-1.26
0.90	17.2	0.32	-10.36	-0.65
1.35	8.1	0.51	-11.16	-0.22
1.80	0.8	0.54	-11.69	0.04
2.25	-4.6	0.49	-11.99	0.18
2.70	-8.2	0.40	-12.11	0.23
3.15	-10.5	0.29	-12.10	0.23
3.60	-11.6	0.20	-12.01	0.20
4.05	-12.1	0.12	-11.88	0.15
4.50	-12.1	0.06	-11.73	0.11
4.95	-11.9	0.02	-11.58	0.07
5.40	-11.6	-0.01	-11.46	0.04
6.75	-10.9	-0.02	-11.20	-0.01
8.10	-10.8	-0.01	-11.10	-0.01

TABLE 21 A

DIA. m		RISE m		RADIUS m		ALPHA deg.
21.500		2.681		22.895		28.00

h1 mm	h2 mm	S m		b mm		d mm
90	200	3.644		280		800

y mm	db mm	qd kN/sq.m		qb kN/sq.m		p m
224	573	2.35		6.18		22.380

A sq.m	qD kN/sq.m	ft N/sq.mm	fb N/sq.mm	Vol. cu.m
386	2.869	0.194	0.399	46.9

s m	Nt kN/m	Mf kNm/m	Nf kN/m	Qf kN/m
0.00	39.2	-0.87	-8.13	-2.14
0.45	29.0	-0.10	-9.52	-1.31
0.90	18.4	0.34	-10.64	-0.68
1.35	9.0	0.54	-11.47	-0.24
1.80	1.3	0.58	-12.02	0.03
2.25	-4.4	0.52	-12.34	0.18
2.70	-8.3	0.43	-12.46	0.24
3.15	-10.6	0.32	-12.46	0.24
3.60	-11.9	0.21	-12.37	0.21
4.05	-12.4	0.13	-12.24	0.16
4.50	-12.5	0.07	-12.09	0.12
4.95	-12.3	0.02	-11.94	0.08
5.40	-12.0	0.00	-11.81	0.05
6.75	-11.3	-0.02	-11.53	0.00
8.10	-11.1	-0.01	-11.43	-0.01

TABLE 22 A			
DIA.	RISE	RADIUS	ALPHA
m	m	m	deg.
22.100	2.755	23.534	28.00

h1	h2	S	b	d
mm	mm	m	mm	mm
90	200	3.695	280	800

y	db	qd	qb	p
mm	mm	kN/sq.m	kN/sq.m	m
224	573	2.34	5.85	23.005

A	qD	ft	fb	Vol.
sq.m	kN/sq.m	N/sq.mm	N/sq.mm	cu.m
408	2.860	0.205	0.420	49.4

s	Nt	Mf	Nf	Qf
m	kN/m	kNm/m	kN/m	kN/m
0.00	41.4	-0.89	-8.35	-2.21
0.45	30.8	-0.10	-9.78	-1.35
0.90	19.8	0.36	-10.93	-0.70
1.35	9.9	0.57	-11.78	-0.25
1.80	1.8	0.61	-12.35	0.03
2.25	-4.1	0.56	-12.68	0.19
2.70	-8.3	0.46	-12.82	0.25
3.15	-10.8	0.34	-12.82	0.25
3.60	-12.2	0.24	-12.74	0.22
4.05	-12.8	0.15	-12.60	0.18
4.50	-12.9	0.08	-12.45	0.13
4.95	-12.7	0.03	-12.30	0.09
5.40	-12.4	0.00	-12.16	0.05
6.75	-11.6	-0.03	-11.87	0.00
8.55	-11.4	-0.01	-11.74	-0.01

TABLE 23 A

DIA.	RISE	RADIUS	ALPHA
m	m	m	deg.
22.800	2.843	24.280	28.00

h1	h2	S	b	d
mm	mm	m	mm	mm
90	200	3.753	280	800

y	db	qd	qb	p
mm	mm	kN/sq.m	kN/sq.m	m
224	573	2.34	5.50	23.734

A	qD	ft	fb	Vol.
sq.m	kN/sq.m	N/sq.mm	N/sq.mm	cu.m
434	2.850	0.218	0.446	52.4

s	Nt	Mf	Nf	Qf
m	kN/m	kNm/m	kN/m	kN/m
0.00	44.1	-0.92	-8.60	-2.28
0.45	32.9	-0.09	-10.07	-1.41
0.90	21.4	0.38	-11.26	-0.74
1.35	11.0	0.60	-12.14	-0.27
1.80	2.5	0.65	-12.74	0.03
2.25	-3.8	0.60	-13.08	0.19
2.70	-8.2	0.49	-13.23	0.26
3.15	-11.0	0.37	-13.25	0.26
3.60	-12.5	0.26	-13.16	0.24
4.05	-13.2	0.16	-13.03	0.19
4.50	-13.3	0.09	-12.87	0.14
4.95	-13.2	0.04	-12.72	0.10
5.40	-12.8	0.00	-12.57	0.06
5.85	-12.5	-0.02	-12.45	0.03
7.20	-11.8	-0.03	-12.21	-0.01
8.55	-11.8	-0.01	-12.12	-0.01

```
TABLE  24  A
 DIA.          RISE         RADIUS         ALPHA
  m             m             m            deg.
23.500        2.930        25.025         28.00
 h1           h2            S             b             d
 mm           mm            m             mm            mm
 90           200          3.810          280           800
 y            db            qd            qb            p
 mm           mm          kN/sq.m       kN/sq.m         m
 224          573          2.34           5.17        24.462
 A            qD            ft            fb            Vol.
sq.m        kN/sq.m      N/sq.mm       N/sq.mm        cu.m
 461         2.840         0.232         0.472          55.4
  s           Nt            Mf            Nf            Qf
  m          kN/m         kNm/m          kN/m          kN/m
0.00         46.9         -0.94         -8.85         -2.36
0.45         35.1         -0.09        -10.37         -1.46
0.90         23.0          0.41        -11.59         -0.77
1.35         12.1          0.64        -12.50         -0.29
1.80          3.2          0.69        -13.12          0.02
2.25         -3.5          0.64        -13.48          0.19
2.70         -8.2          0.53        -13.65          0.27
3.15        -11.2          0.41        -13.67          0.28
3.60        -12.9          0.29        -13.59          0.25
4.05        -13.6          0.18        -13.46          0.20
4.50        -13.8          0.10        -13.30          0.15
4.95        -13.6          0.05        -13.14          0.11
5.40        -13.3          0.01        -12.99          0.07
5.85        -12.9         -0.02        -12.86          0.04
7.20        -12.2         -0.03        -12.60         -0.01
9.00        -12.1         -0.01        -12.50         -0.01
```

TABLE 25 A

DIA.		RISE		RADIUS		ALPHA
m		m		m		deg.
24.000		2.992		25.558		28.00

h1	h2	S	b	d
mm	mm	m	mm	mm
100	250	4.230	300	1000

y	db	qd	qb	p
mm	mm	kN/sq.m	kN/sq.m	m
273	717	2.68	6.12	24.983

A	qD	ft	fb	Vol.
sq.m	kN/sq.m	N/sq.mm	N/sq.mm	cu.m
481	3.411	0.170	0.360	69.5

s	Nt	Mf	Nf	Qf
m	kN/m	kNm/m	kN/m	kN/m
0.00	43.0	-1.25	-8.84	-2.52
0.45	33.8	-0.31	-10.22	-1.68
0.90	23.9	0.28	-11.40	-1.00
1.35	14.5	0.62	-12.33	-0.50
1.80	6.3	0.76	-13.03	-0.14
2.25	-0.3	0.77	-13.50	0.08
2.70	-5.4	0.70	-13.79	0.21
3.15	-9.0	0.58	-13.92	0.27
3.60	-11.4	0.46	-13.94	0.28
4.05	-12.9	0.34	-13.88	0.26
4.50	-13.7	0.23	-13.77	0.22
4.95	-14.0	0.14	-13.62	0.17
5.40	-13.9	0.08	-13.47	0.13
5.85	-13.8	0.03	-13.33	0.09
7.65	-12.8	-0.03	-12.90	0.00
9.00	-12.5	-0.02	-12.75	-0.01

```
TABLE  26  A
 DIA.          RISE         RADIUS        ALPHA
   m             m             m           deg.
 24.700        3.080        26.303        28.00
  h1          h2         S          b           d
  mm          mm         m          mm          mm
 100         250       4.291       300        1000
   y          db         qd          qb          p
  mm          mm      kN/sq.m     kN/sq.m        m
 273         717        2.67        5.78       25.712
   A          qD         ft          fb         Vol.
 sq.m      kN/sq.m    N/sq.mm     N/sq.mm      cu.m
 509        3.398      0.180       0.380        73.3
   s          Nt         Mf          Nf          Qf
   m         kN/m      kNm/m        kN/m        kN/m
 0.00        45.6      -1.28       -9.09       -2.60
 0.45        35.9      -0.31      -10.51       -1.74
 0.90        25.5       0.31      -11.72       -1.05
 1.35        15.7       0.65      -12.68       -0.52
 1.80         7.2       0.80      -13.40       -0.16
 2.25         0.2       0.81      -13.89        0.08
 2.70        -5.1       0.74      -14.19        0.22
 3.15        -9.0       0.63      -14.34        0.28
 3.60       -11.6       0.50      -14.36        0.29
 4.05       -13.2       0.37      -14.30        0.27
 4.50       -14.1       0.26      -14.19        0.23
 4.95       -14.4       0.16      -14.05        0.18
 5.40       -14.4       0.09      -13.90        0.14
 5.85       -14.2       0.04      -13.75        0.10
 6.30       -13.9       0.00      -13.61        0.06
 7.65       -13.2      -0.04      -13.29        0.00
 9.45       -12.8      -0.02      -13.11       -0.01
```

TABLE 27 A

DIA.		RISE		RADIUS		ALPHA
m		m		m		deg.
25.500		3.179		27.155		28.00

h1	h2	S		b	d
mm	mm	m		mm	mm
100	250	4.360		300	1000

y	db	qd		qb	p
mm	mm	kN/sq.m		kN/sq.m	m
273	717	2.67		5.42	26.544

A	qD	ft	fb	Vol.
sq.m	kN/sq.m	N/sq.mm	N/sq.mm	cu.m
543	3.383	0.192	0.404	77.8

s	Nt	Mf	Nf	Qf
m	kN/m	kNm/m	kN/m	kN/m
0.00	48.6	-1.32	-9.37	-2.69
0.45	38.4	-0.31	-10.83	-1.80
0.90	27.5	0.33	-12.08	-1.09
1.35	17.2	0.69	-13.07	-0.55
1.80	8.2	0.85	-13.82	-0.17
2.25	0.9	0.87	-14.33	0.08
2.70	-4.8	0.80	-14.65	0.22
3.15	-8.9	0.68	-14.81	0.29
3.60	-11.8	0.54	-14.84	0.31
4.05	-13.5	0.41	-14.79	0.29
4.50	-14.5	0.29	-14.68	0.25
4.95	-14.9	0.19	-14.54	0.20
5.40	-15.0	0.11	-14.38	0.15
5.85	-14.8	0.05	-14.22	0.11
6.30	-14.5	0.01	-14.08	0.07
8.10	-13.4	-0.04	-13.68	0.00
9.90	-13.2	-0.02	-13.53	-0.01

TABLE 28 A

DIA.		RISE		RADIUS		ALPHA
m		m		m		deg.
26.200		3.267		27.900		28.00
h1	h2	S		b		d
mm	mm	m		mm		mm
100	250	4.419		300		1000
y	db	qd		qb		p
mm	mm	kN/sq.m		kN/sq.m		m
273	717	2.66		5.14		27.273
A	qD	ft		fb		Vol.
sq.m	kN/sq.m	N/sq.mm		N/sq.mm		cu.m
573	3.371	0.203		0.425		81.8

s	Nt	Mf	Nf	Qf
m	kN/m	kNm/m	kN/m	kN/m
0.00	51.3	-1.35	-9.62	-2.77
0.45	40.6	-0.31	-11.12	-1.86
0.90	29.3	0.35	-12.40	-1.13
1.35	18.6	0.73	-13.42	-0.58
1.80	9.2	0.90	-14.19	-0.19
2.25	1.5	0.92	-14.72	0.07
2.70	-4.5	0.85	-15.05	0.23
3.15	-8.9	0.73	-15.22	0.30
3.60	-11.9	0.58	-15.26	0.32
4.05	-13.8	0.44	-15.21	0.30
4.50	-14.9	0.32	-15.11	0.26
4.95	-15.4	0.21	-14.96	0.21
5.40	-15.4	0.12	-14.80	0.17
5.85	-15.2	0.06	-14.64	0.12
6.30	-14.9	0.01	-14.49	0.08
6.75	-14.6	-0.02	-14.36	0.05
8.10	-13.8	-0.04	-14.07	0.00
9.90	-13.6	-0.02	-13.92	-0.01

TABLE 29 A

DIA.	RISE	RADIUS	ALPHA
m	m	m	deg.
27.000	3.366	28.752	28.00

h1	h2	S	b	d
mm	mm	m	mm	mm
100	250	4.486	300	1000

y	db	qd	qb	p
mm	mm	kN/sq.m	kN/sq.m	m
273	717	2.66	4.84	28.106

A	qD	ft	fb	Vol.
sq.m	kN/sq.m	N/sq.mm	N/sq.mm	cu.m
608	3.358	0.215	0.450	86.5

s	Nt	Mf	Nf	Qf
m	kN/m	kNm/m	kN/m	kN/m
0.00	54.5	-1.38	-9.90	-2.86
0.45	43.3	-0.31	-11.44	-1.93
0.90	31.4	0.38	-12.76	-1.18
1.35	20.2	0.78	-13.81	-0.61
1.80	10.4	0.95	-14.60	-0.20
2.25	2.2	0.98	-15.16	0.07
2.70	-4.1	0.91	-15.51	0.23
3.15	-8.8	0.78	-15.69	0.31
3.60	-12.0	0.63	-15.74	0.34
4.05	-14.1	0.49	-15.70	0.32
4.50	-15.3	0.35	-15.59	0.28
4.95	-15.9	0.23	-15.45	0.23
5.40	-16.0	0.14	-15.29	0.18
5.85	-15.8	0.07	-15.12	0.13
6.30	-15.5	0.02	-14.97	0.09
6.75	-15.1	-0.01	-14.83	0.06
8.55	-14.1	-0.04	-14.46	-0.01
10.35	-14.0	-0.02	-14.34	-0.01

TABLE 30 A

DIA. m		RISE m		RADIUS m		ALPHA deg.
27.800		3.466		29.604		28.00

h1 mm	h2 mm	S m	b mm	d mm
100	250	4.552	300	1000

y mm	db mm	qd kN/sq.m	qb kN/sq.m	p m
273	717	2.66	4.56	28.938

A sq.m	qD kN/sq.m	ft N/sq.mm	fb N/sq.mm	Vol. cu.m
645	3.345	0.228	0.476	91.4

s m	Nt kN/m	Mf kNm/m	Nf kN/m	Qf kN/m
0.00	57.8	-1.42	-10.18	-2.95
0.45	46.0	-0.31	-11.76	-2.00
0.90	33.6	0.41	-13.12	-1.23
1.35	21.9	0.82	-14.20	-0.64
1.80	11.6	1.01	-15.02	-0.22
2.25	3.0	1.04	-15.60	0.07
2.70	-3.7	0.97	-15.97	0.24
3.15	-8.6	0.84	-16.16	0.32
3.60	-12.1	0.69	-16.23	0.35
4.05	-14.4	0.53	-16.19	0.34
4.50	-15.7	0.39	-16.08	0.30
4.95	-16.4	0.26	-15.94	0.25
5.40	-16.5	0.16	-15.77	0.20
5.85	-16.4	0.09	-15.61	0.15
6.30	-16.1	0.03	-15.45	0.10
6.75	-15.7	-0.01	-15.30	0.07
7.20	-15.3	-0.03	-15.18	0.04
9.00	-14.4	-0.04	-14.86	-0.01
10.80	-14.4	-0.01	-14.77	-0.01

TABLE 31 A

DIA. m	RISE m	RADIUS m	ALPHA deg.
28.700	3.578	30.563	28.00

h1 mm	h2 mm	S m	b mm	d mm
100	250	4.625	300	1000

y mm	db mm	qd kN/sq.m	qb kN/sq.m	p m
273	717	2.65	4.28	29.875

A sq.m	qD kN/sq.m	ft N/sq.mm	fb N/sq.mm	Vol. cu.m
687	3.331	0.243	0.505	97.0

s m	Nt kN/m	Mf kNm/m	Nf kN/m	Qf kN/m
0.00	61.6	-1.45	-10.49	-3.05
0.45	49.2	-0.31	-12.12	-2.08
0.90	36.1	0.44	-13.52	-1.29
1.35	23.8	0.88	-14.64	-0.68
1.80	13.0	1.08	-15.49	-0.24
2.25	4.0	1.11	-16.10	0.06
2.70	-3.1	1.04	-16.48	0.24
3.15	-8.5	0.91	-16.70	0.34
3.60	-12.2	0.75	-16.77	0.37
4.05	-14.7	0.58	-16.74	0.36
4.50	-16.2	0.43	-16.63	0.32
4.95	-16.9	0.30	-16.49	0.27
5.40	-17.1	0.19	-16.32	0.21
5.85	-17.0	0.10	-16.15	0.16
6.30	-16.7	0.04	-15.99	0.11
6.75	-16.3	0.00	-15.83	0.08
7.20	-15.9	-0.03	-15.70	0.04
9.00	-14.9	-0.04	-15.36	-0.01
10.80	-14.8	-0.02	-15.26	-0.01

TABLE 32 A

DIA.	RISE	RADIUS	ALPHA
m	m	m	deg.
29.500	3.678	31.415	28.00
h1	h2	S	b	d
mm	mm	m	mm	mm
100	250	4.689	300	1000
y	db	qd	qb	p
mm	mm	kN/sq.m	kN/sq.m	m
273	717	2.65	4.05	30.708
A	qD	ft	fb	Vol.
sq.m	kN/sq.m	N/sq.mm	N/sq.mm	cu.m
726	3.319	0.257	0.532	102.1

s	Nt	Mf	Nf	Qf
m	kN/m	kNm/m	kN/m	kN/m
0.00	65.1	-1.49	-10.76	-3.15
0.45	52.1	-0.30	-12.44	-2.15
0.90	38.5	0.47	-13.87	-1.34
1.35	25.6	0.93	-15.03	-0.71
1.80	14.3	1.14	-15.91	-0.25
2.25	4.9	1.18	-16.54	0.06
2.70	-2.6	1.11	-16.94	0.25
3.15	-8.3	0.97	-17.17	0.35
3.60	-12.3	0.80	-17.25	0.38
4.05	-15.0	0.63	-17.22	0.37
4.50	-16.6	0.47	-17.12	0.34
4.95	-17.4	0.33	-16.98	0.29
5.40	-17.7	0.21	-16.81	0.23
5.85	-17.6	0.12	-16.64	0.18
6.30	-17.3	0.05	-16.47	0.13
6.75	-16.9	0.01	-16.31	0.09
7.20	-16.4	-0.03	-16.17	0.05
7.65	-16.1	-0.04	-16.05	0.03
9.45	-15.2	-0.04	-15.77	-0.02
11.25	-15.2	-0.01	-15.68	-0.01

TABLE 33 A

DIA.		RISE		RADIUS		ALPHA
m		m		m		deg.
30.000		3.740		31.947		28.00
h1	h2		S		b	d
mm	mm		m		mm	mm
130	290		5.180		300	1200
y	db	qd		qb		p
mm	mm	kN/sq.m		kN/sq.m		m
345	872	3.40		6.62		31.229
A	qD		ft		fb	Vol.
sq.m	kN/sq.m		N/sq.mm		N/sq.mm	cu.m
751	4.169		0.210		0.435	132.7

s	Nt	Mf	Nf	Qf
m	kN/m	kNm/m	kN/m	kN/m
0.00	61.6	-2.02	-10.57	-3.40
0.45	51.0	-0.71	-12.15	-2.44
0.90	39.6	0.21	-13.54	-1.64
1.35	28.3	0.79	-14.71	-1.00
1.80	18.0	1.13	-15.66	-0.50
2.25	8.9	1.27	-16.38	-0.14
2.70	1.4	1.27	-16.90	0.11
3.15	-4.6	1.18	-17.24	0.26
3.60	-9.3	1.04	-17.44	0.35
4.05	-12.7	0.88	-17.51	0.38
4.50	-15.0	0.70	-17.50	0.37
4.95	-16.5	0.54	-17.42	0.34
5.40	-17.4	0.40	-17.29	0.30
5.85	-17.8	0.27	-17.14	0.25
6.30	-17.8	0.17	-16.98	0.20
6.75	-17.6	0.09	-16.82	0.15
7.20	-17.3	0.04	-16.66	0.11
7.65	-17.0	-0.01	-16.52	0.07
9.45	-15.9	-0.06	-16.10	0.00
11.70	-15.6	-0.02	-15.91	-0.02

```
TABLE 34 A
```

DIA.	RISE	RADIUS	ALPHA
m	m	m	deg.
30.900	3.853	32.905	28.00

h1	h2	S	b	d
mm	mm	m	mm	mm
130	290	5.257	300	1200

y	db	qd	qb	p
mm	mm	kN/sq.	kN/sq.m	m
345	872	3.39	6.24	32.165

A	qD	ft	fb	Vol.
sq.m	kN/sq.m	N/sq.mm	N/sq.mm	cu.m
797	4.155	0.223	0.460	140.3

s	Nt	Mf	Nf	Qf
m	kN/m	kNm/m	kN/m	kN/m
0.00	65.4	-2.07	-10.88	-3.51
0.45	54.2	-0.72	-12.50	-2.53
0.90	42.2	0.23	-13.93	-1.71
1.35	30.4	0.84	-15.13	-1.04
1.80	19.6	1.19	-16.11	-0.53
2.25	10.2	1.34	-16.85	-0.16
2.70	2.2	1.35	-17.40	0.10
3.15	-4.2	1.26	-17.76	0.27
3.60	-9.1	1.12	-17.97	0.36
4.05	-12.7	0.95	-18.05	0.40
4.50	-15.3	0.77	-18.05	0.39
4.95	-16.9	0.59	-17.97	0.36
5.40	-17.9	0.44	-17.85	0.32
5.85	-18.3	0.31	-17.69	0.27
6.30	-18.4	0.20	-17.53	0.22
6.75	-18.3	0.11	-17.36	0.17
7.20	-18.0	0.05	-17.20	0.12
7.65	-17.6	0.00	-17.05	0.08
9.90	-16.2	-0.06	-16.55	-0.01
11.70	-16.0	-0.03	-16.40	-0.02

TABLE 35 A

DIA.	RISE	RADIUS	ALPHA
m	m	m	deg.
31.800	3.965	33.864	28.00

h1	h2	S	b	d
mm	mm	m	mm	mm
130	290	5.333	300	1200

y	db	qd	qb	p
mm	mm	kN/sq.m	kN/sq.m	m
345	872	3.39	5.89	33.102

A	qD	ft	fb	Vol.
sq.m	kN/sq.m	N/sq.mm	N/sq.mm	cu.m
844	4.141	0.236	0.486	148.0

s	Nt	Mf	Nf	Qf
m	kN/m	kNm/m	kN/m	kN/m
0.00	69.2	-2.12	-11.18	-3.62
0.45	57.5	-0.72	-12.85	-2.62
0.90	45.0	0.26	-14.31	-1.77
1.35	32.7	0.90	-15.55	-1.09
1.80	21.4	1.26	-16.56	-0.56
2.25	11.5	1.42	-17.33	-0.17
2.70	3.1	1.43	-17.89	0.10
3.15	-3.6	1.35	-18.27	0.28
3.60	-8.9	1.20	-18.49	0.37
4.05	-12.7	1.02	-18.59	0.41
4.50	-15.5	0.83	-18.59	0.41
4.95	-17.3	0.65	-18.52	0.39
5.40	-18.4	0.49	-18.40	0.34
5.85	-18.9	0.34	-18.24	0.29
6.30	-19.0	0.23	-18.08	0.23
6.75	-18.9	0.13	-17.90	0.18
7.20	-18.6	0.06	-17.74	0.14
7.65	-18.2	0.01	-17.58	0.10
8.10	-17.9	-0.03	-17.44	0.06
9.90	-16.7	-0.06	-17.05	-0.01
12.15	-16.4	-0.02	-16.88	-0.02

```
TABLE  36 A
 DIA.          RISE          RADIUS          ALPHA
  m             m              m             deg.
32.800        4.090         34.929          28.00
 h1           h2        S         b             d
 mm           mm        m         mm            mm
 130          290      5.417      300          1200
  y           db        qd        qb            p
 mm           mm     kN/sq.m   kN/sq.m          m
 345          872      3.39      5.54         34.143
  A           qD        ft        fb           Vol.
sq.m       kN/sq.m   N/sq.mm   N/sq.mm         cu.m
 898         4.126    0.251     0.515          157.0
  s           Nt        Mf        Nf            Qf
  m          kN/m      kNm/m     kN/m          kN/m
 0.00        73.6     -2.18     -11.51         -3.74
 0.45        61.3     -0.73     -13.23         -2.71
 0.90        48.1      0.29     -14.74         -1.84
 1.35        35.2      0.96     -16.02         -1.14
 1.80        23.4      1.34     -17.05         -0.60
 2.25        13.0      1.51     -17.85         -0.19
 2.70         4.2      1.53     -18.44          0.10
 3.15        -3.0      1.44     -18.84          0.28
 3.60        -8.6      1.29     -19.08          0.39
 4.05       -12.7      1.10     -19.19          0.43
 4.50       -15.7      0.91     -19.20          0.44
 4.95       -17.7      0.72     -19.13          0.41
 5.40       -18.9      0.54     -19.01          0.36
 5.85       -19.5      0.39     -18.86          0.31
 6.30       -19.7      0.26     -18.69          0.26
 6.75       -19.6      0.16     -18.51          0.20
 7.20       -19.3      0.08     -18.34          0.15
 7.65       -19.0      0.02     -18.18          0.11
 8.10       -18.5     -0.02     -18.03          0.07
10.35       -17.1     -0.06     -17.56         -0.01
12.60       -16.9     -0.02     -17.42         -0.02
```

TABLE 37 A

DIA.	RISE		RADIUS	ALPHA
m	m		m	deg.
33.800	4.214		35.994	28.00
h1	h2	S	b	d
mm	mm	m	mm	mm
130	290	5.499	300	1200
y	db	qd	qb	p
mm	mm	kN/sq.m	kN/sq.m	m
345	872	3.38	5.22	35.184
A	qD	ft	fb	Vol.
sq.m	kN/sq.m	N/sq.mm	N/sq.mm	cu.m
953	4.112	0.266	0.545	166.1
s	Nt	Mf	Nf	Qf
m	kN/m	kNm/m	kN/m	kN/m
0.00	78.1	-2.23	-11.84	-3.87
0.45	65.2	-0.73	-13.61	-2.81
0.90	51.4	0.32	-15.16	-1.92
1.35	37.9	1.02	-16.48	-1.19
1.80	25.5	1.42	-17.55	-0.63
2.25	14.6	1.61	-18.38	-0.21
2.70	5.3	1.63	-18.99	0.09
3.15	-2.3	1.54	-19.41	0.28
3.60	-8.2	1.38	-19.67	0.40
4.05	-12.7	1.19	-19.79	0.45
4.50	-15.9	0.99	-19.81	0.46
4.95	-18.1	0.78	-19.75	0.43
5.40	-19.5	0.60	-19.63	0.39
5.85	-20.2	0.44	-19.47	0.33
6.30	-20.4	0.30	-19.30	0.28
6.75	-20.3	0.19	-19.12	0.22
7.20	-20.1	0.10	-18.94	0.17
7.65	-19.7	0.03	-18.78	0.12
8.10	-19.2	-0.01	-18.62	0.08
8.55	-18.8	-0.04	-18.49	0.05
10.80	-17.5	-0.06	-18.07	-0.02
13.05	-17.4	-0.02	-17.95	-0.01

```
TABLE 38 A
 DIA.          RISE          RADIUS        ALPHA
  m             m              m            deg.
34.800        4.339         37.059        28.00
 h1            h2       S        b           d
 mm            mm       m        mm          mm
 130          290      5.579    300        1200
  y            db      qd         qb          p
 mm            mm    kN/sq.m    kN/sq.m      m
 345          872     3.38       4.92      36.225
  A            qD       ft         fb        Vol.
sq.m        kN/sq.m  N/sq.mm    N/sq.mm     cu.m
1010         4.099    0.282      0.576      175.5
  s           Nt       Mf         Nf          Qf
  m          kN/m     kNm/m      kN/m        kN/m
 0.00        82.8    -2.29      -12.17      -3.99
 0.45        69.2    -0.74      -13.99      -2.91
 0.90        54.8     0.36      -15.59      -1.99
 1.35        40.7     1.08      -16.94      -1.25
 1.80        27.7     1.51      -18.04      -0.67
 2.25        16.2     1.70      -18.90      -0.23
 2.70         6.5     1.73      -19.54       0.08
 3.15        -1.5     1.64      -19.98       0.29
 3.60        -7.8     1.48      -20.25       0.41
 4.05       -12.6     1.28      -20.39       0.47
 4.50       -16.1     1.07      -20.41       0.48
 4.95       -18.5     0.86      -20.36       0.46
 5.40       -20.0     0.66      -20.24       0.41
 5.85       -20.8     0.49      -20.09       0.36
 6.30       -21.1     0.34      -19.92       0.30
 6.75       -21.1     0.22      -19.73       0.24
 7.20       -20.8     0.12      -19.55       0.19
 7.65       -20.4     0.05      -19.38       0.14
 8.10       -20.0     0.00      -19.22       0.10
 8.55       -19.5    -0.04      -19.07       0.06
 9.00       -19.1    -0.06      -18.95       0.03
11.25       -17.9    -0.06      -18.58      -0.02
13.50       -18.0    -0.02      -18.49      -0.01
```

TABLE 39 A

DIA.	RISE	RADIUS	ALPHA
m	m	m	deg.
35.000	4.364	37.272	28.00

h1	h2	S	b	d
mm	mm	m	mm	mm
130	290	5.595	300	1200

y	db	qd	qb	p
mm	mm	kN/sq.m	kN/sq.m	m
345	872	3.38	4.87	36.433

A	qD	ft	fb	Vol.
sq.m	kN/sq.m	N/sq.mm	N/sq.mm	cu.m
1022	4.096	0.285	0.582	177.4

s	Nt	Mf	Nf	Qf
m	kN/m	kNm/m	kN/m	kN/m
0.00	83.7	-2.30	-12.24	-4.02
0.45	70.0	-0.74	-14.06	-2.93
0.90	55.4	0.37	-15.67	-2.01
1.35	41.2	1.09	-17.03	-1.26
1.80	28.1	1.52	-18.14	-0.67
2.25	16.6	1.72	-19.01	-0.23
2.70	6.7	1.75	-19.65	0.08
3.15	-1.4	1.66	-20.10	0.29
3.60	-7.7	1.50	-20.37	0.41
4.05	-12.6	1.30	-20.51	0.47
4.50	-16.1	1.08	-20.54	0.48
4.95	-18.5	0.87	-20.48	0.46
5.40	-20.1	0.67	-20.37	0.42
5.85	-20.9	0.50	-20.21	0.36
6.30	-21.3	0.35	-20.04	0.30
6.75	-21.2	0.22	-19.85	0.24
7.20	-21.0	0.12	-19.67	0.19
7.65	-20.6	0.05	-19.50	0.14
8.10	-20.1	0.00	-19.34	0.10
8.55	-19.6	-0.04	-19.19	0.06
9.00	-19.2	-0.06	-19.07	0.04
11.25	-18.0	-0.06	-18.69	-0.02
13.50	-18.1	-0.02	-18.60	-0.01

```
TABLE  40 A
  DIA.          RISE          RADIUS        ALPHA
   m              m              m           deg.
 36.100        4.501         38.443         28.00
  h1           h2        S          b           d
  mm           mm        m         mm          mm
 130          290      5.683      300        1200
  y           db       qd          qb          p
  mm          mm     kN/sq.m     kN/sq.m       m
 345         872      3.37        4.57       37.578
  A           qD       ft          fb         Vol.
 sq.m       kN/sq.m  N/sq.mm     N/sq.mm     cu.m
 1087        4.082    0.303       0.617      188.1
  s           Nt       Mf          Nf          Qf
  m          kN/m     kNm/m       kN/m        kN/m
 0.00        89.0    -2.35       -12.60      -4.16
 0.45        74.5    -0.74       -14.48      -3.04
 0.90        59.3     0.41       -16.13      -2.09
 1.35        44.4     1.17       -17.54      -1.32
 1.80        30.7     1.62       -18.68      -0.71
 2.25        18.5     1.83       -19.58      -0.26
 2.70         8.1     1.87       -20.25       0.07
 3.15        -0.5     1.78       -20.72       0.29
 3.60        -7.2     1.62       -21.02       0.43
 4.05       -12.5     1.41       -21.17       0.49
 4.50       -16.3     1.18       -21.21       0.51
 4.95       -18.9     0.95       -21.16       0.49
 5.40       -20.6     0.74       -21.05       0.45
 5.85       -21.6     0.55       -20.89       0.39
 6.30       -22.0     0.39       -20.72       0.33
 6.75       -22.0     0.26       -20.53       0.27
 7.20       -21.8     0.15       -20.34       0.21
 7.65       -21.4     0.07       -20.16       0.16
 8.10       -20.9     0.01       -19.99       0.11
 8.55       -20.4    -0.03       -19.84       0.07
 9.00       -19.9    -0.06       -19.71       0.04
11.70       -18.5    -0.06       -19.27      -0.02
13.95       -18.6    -0.01       -19.19      -0.01
```

Tables 1B to 40B   For Type B Domes

TABLE 1 B

DIA.	RISE	RADIUS	ALPHA
m	m	m	deg.
12.000	1.496	12.779	28.00

h1	h2	S	b	d
mm	mm	m	mm	mm
75	180	2.553	250	700

y	db	qd	qb	p
mm	mm	kN/sq.m	kN/sq.m	m
-41	266	2.03	13.78	12.491

A	qD	ft	fb	Vol.
sq.m	kN/sq.m	N/sq.mm	N/sq.mm	cu.m
120	2.636	-0.001	0.266	13.4

s	Nt	Mf	Nf	Qf
m	kN/m	kNm/m	kN/m	kN/m
0.00	21.7	0.10	-4.91	-1.00
0.45	10.3	0.38	-6.15	-0.28
0.90	1.8	0.41	-6.89	0.09
1.35	-3.6	0.33	-7.21	0.24
1.80	-6.4	0.22	-7.25	0.24
2.25	-7.5	0.12	-7.13	0.19
2.70	-7.6	0.05	-6.94	0.12
3.60	-6.9	-0.01	-6.58	0.03
4.50	-6.4	-0.02	-6.36	-0.01

TABLE 2 B

DIA.		RISE		RADIUS		ALPHA
m		m		m		deg.
12.400		1.546		13.205		28.00

h1	h2	S	b	d
mm	mm	m	mm	mm
75	180	2.595	250	700

y	db	qd	qb	p
mm	mm	kN/sq.m	kN/sq.m	m
-41	266	2.03	12.90	12.908

A	qD	ft	fb	Vol.
sq.m	kN/sq.m	N/sq.mm	N/sq.mm	cu.m
128	2.624	0.001	0.281	14.3

s	Nt	Mf	Nf	Qf
m	kN/m	kNm/m	kN/m	kN/m
0.00	23.1	0.12	-5.06	-1.04
0.45	11.2	0.40	-6.35	-0.30
0.90	2.3	0.44	-7.11	0.09
1.35	-3.5	0.35	-7.46	0.24
1.80	-6.5	0.24	-7.51	0.26
2.25	-7.7	0.13	-7.39	0.20
2.70	-7.9	0.06	-7.19	0.13
3.15	-7.6	0.01	-6.99	0.07
3.60	-7.2	-0.01	-6.82	0.03
4.50	-6.6	-0.02	-6.59	-0.01

TABLE 3 B

DIA.		RISE		RADIUS		ALPHA
m		m		m		deg.
12.700		1.583		13.524		28.00

h1	h2	S	b	d
mm	mm	m	mm	mm
75	180	2.626	250	700

y	db	qd	qb	p
mm	mm	kN/sq.m	kN/sq.m	m
-41	266	2.02	12.30	13.220

A	qD	ft	fb	Vol.
sq.m	kN/sq.m	N/sq.mm	N/sq.mm	cu.m
135	2.615	0.004	0.292	14.9

s	Nt	Mf	Nf	Qf
m	kN/m	kNm/m	kN/m	kN/m
0.00	24.2	0.13	-5.18	-1.07
0.45	11.9	0.42	-6.49	-0.31
0.90	2.6	0.46	-7.28	0.09
1.35	-3.4	0.37	-7.64	0.25
1.80	-6.6	0.25	-7.69	0.27
2.25	-7.9	0.14	-7.58	0.21
2.70	-8.1	0.06	-7.38	0.14
3.15	-7.9	0.01	-7.18	0.08
4.05	-7.1	-0.02	-6.86	0.01
4.95	-6.7	-0.01	-6.69	-0.01

TABLE 4 B

DIA.		RISE		RADIUS		ALPHA
m		m		m		deg.
13.100		1.633		13.950		28.00

h1	h2	S	b	d
mm	mm	m	mm	mm
75	180	2.667	250	700

y	db	qd	qb	p
mm	mm	kN/sq.m	kN/sq.m	m
-41	266	2.02	11.56	13.636

A	qD	ft	fb	Vol.
sq.m	kN/sq.m	N/sq.mm	N/sq.mm	cu.m
143	2.604	0.007	0.307	15.8

s	Nt	Mf	Nf	Qf
m	kN/m	kNm/m	kN/m	kN/m
0.00	25.6	0.14	-5.34	-1.10
0.45	12.8	0.45	-6.69	-0.33
0.90	3.1	0.49	-7.50	0.09
1.35	-3.2	0.40	-7.88	0.26
1.80	-6.7	0.27	-7.95	0.28
2.25	-8.2	0.16	-7.83	0.23
2.70	-8.4	0.07	-7.64	0.15
3.15	-8.2	0.02	-7.43	0.09
4.05	-7.3	-0.02	-7.09	0.01
4.95	-6.9	-0.02	-6.92	-0.01

116

TABLE 5 B

DIA. m	RISE m	RADIUS m	ALPHA deg.
13.500	1.683	14.376	28.00

h1 mm	h2 mm	S m	b mm	d mm
75	180	2.708	250	700

y mm	db mm	qd kN/sq.m	qb kN/sq.m	p m
-41	266	2.02	10.89	14.053

A sq.m	qD kN/sq.m	ft N/sq.mm	fb N/sq.mm	Vol. cu.m
152	2.594	0.010	0.323	16.7

s m	Nt kN/m	Mf kNm/m	Nf kN/m	Qf kN/m
0.00	27.2	0.15	-5.49	-1.14
0.45	13.8	0.47	-6.88	-0.34
0.90	3.6	0.52	-7.72	0.09
1.35	-3.1	0.43	-8.12	0.27
1.80	-6.8	0.30	-8.20	0.29
2.25	-8.4	0.18	-8.09	0.24
2.70	-8.8	0.08	-7.89	0.17
3.15	-8.5	0.02	-7.68	0.10
4.05	-7.6	-0.02	-7.33	0.01
4.95	-7.1	-0.02	-7.14	-0.01

```
TABLE 6 B
 DIA.         RISE         RADIUS        ALPHA
  m            m             m           deg.
13.900        1.733        14.802        28.00
 h1     h2        S          b            d
 mm     mm        m          mm           mm
 75     180      2.748      250          700
  y     db       qd          qb           p
 mm     mm     kN/sq.m     kN/sq.m        m
 -41    266     2.01        10.27       14.469
  A     qD        ft          fb         Vol.
sq.m  kN/sq.m  N/sq.mm     N/sq.mm       cu.m
 161    2.583    0.013       0.339        17.6
  s      Nt        Mf          Nf          Qf
  m     kN/m     kNm/m        kN/m        kN/m
0.00    28.7     0.17        -5.65       -1.18
0.45    14.8     0.50        -7.07       -0.36
0.90     4.2     0.55        -7.94        0.09
1.35    -2.9     0.46        -8.36        0.28
1.80    -6.9     0.32        -8.45        0.31
2.25    -8.6     0.19        -8.35        0.26
2.70    -9.1     0.09        -8.15        0.18
3.15    -8.8     0.03        -7.93        0.11
3.60    -8.4    -0.01        -7.73        0.05
4.50    -7.5    -0.02        -7.45        0.00
5.40    -7.2    -0.01        -7.32       -0.01
```

TABLE 7 B

DIA.		RISE		RADIUS		ALPHA
m		m		m		deg.
14.300		1.783		15.228		28.00
h1	h2		S		b	d
mm	mm		m		mm	mm
75	180		2.787		250	700
y	db	qd		qb		p
mm	mm	kN/sq.m		kN/sq.m		m
-41	266	2.01		9.70		14.886
A	qD		ft		fb	Vol.
sq.m	kN/sq.m		N/sq.mm		N/sq.mm	cu.m
171	2.573		0.017		0.355	18.6

s	Nt	Mf	Nf	Qf
m	kN/m	kNm/m	kN/m	kN/m
0.00	30.3	0.18	-5.80	-1.22
0.45	15.9	0.52	-7.26	-0.38
0.90	4.7	0.58	-8.16	0.09
1.35	-2.7	0.49	-8.60	0.28
1.80	-6.9	0.35	-8.70	0.32
2.25	-8.9	0.21	-8.60	0.27
2.70	-9.4	0.11	-8.40	0.19
3.15	-9.2	0.04	-8.18	0.12
3.60	-8.7	0.00	-7.97	0.06
4.50	-7.8	-0.03	-7.68	0.00
5.40	-7.4	-0.02	-7.55	-0.01

TABLE 8 B

DIA.		RISE		RADIUS		ALPHA
m		m		m		deg.
14.800		1.845		15.761		28.00
h1	h2		S		b	d
mm	mm		m		mm	mm
75	180		2.835		250	700
y	db		qd		qb	p
mm	mm		kN/sq.m		kN/sq.m	m
-41	266		2.01		9.06	15.406
A	qD		ft		fb	Vol.
sq.m	kN/sq.m		N/sq.mm		N/sq.mm	cu.m
183	2.561		0.022		0.375	19.8
s	Nt		Mf		Nf	Qf
m	kN/m		kNm/m		kN/m	kN/m
0.00	32.3		0.20		-5.99	-1.27
0.45	17.2		0.56		-7.50	-0.40
0.90	5.5		0.62		-8.44	0.08
1.35	-2.4		0.52		-8.90	0.29
1.80	-7.0		0.38		-9.02	0.34
2.25	-9.1		0.24		-8.92	0.29
2.70	-9.8		0.12		-8.72	0.21
3.15	-9.6		0.05		-8.50	0.13
3.60	-9.1		0.00		-8.28	0.07
4.50	-8.1		-0.03		-7.97	0.00
5.40	-7.7		-0.02		-7.82	-0.01

TABLE 9 B

DIA.		RISE		RADIUS		ALPHA
m		m		m		deg.
15.200		1.895		16.186		28.00

h1	h2	S	b	d
mm	mm	m	mm	mm
75	180	2.873	250	700

y	db	qd	qb	p
mm	mm	kN/sq.m	kN/sq.m	m
-41	266	2.00	8.59	15.822

A	qD	ft	fb	Vol.
sq.m	kN/sq.m	N/sq.mm	N/sq.mm	cu.m
193	2.552	0.026	0.392	20.8

s	Nt	Mf	Nf	Qf
m	kN/m	kNm/m	kN/m	kN/m
0.00	34.0	0.21	-6.14	-1.31
0.45	18.3	0.58	-7.69	-0.42
0.90	6.1	0.65	-8.65	0.08
1.35	-2.1	0.56	-9.13	0.30
1.80	-7.0	0.40	-9.27	0.35
2.25	-9.3	0.26	-9.18	0.30
2.70	-10.1	0.14	-8.98	0.23
3.15	-9.9	0.05	-8.75	0.14
3.60	-9.4	0.00	-8.53	0.08
4.50	-8.4	-0.03	-8.20	0.00
5.85	-7.9	-0.01	-8.01	-0.01

TABLE 10 B

DIA. m	RISE m	RADIUS m	ALPHA deg.
15.700	1.957	16.719	28.00

h1 mm	h2 mm	S m	b mm	d mm
75	180	2.920	250	700

y mm	db mm	qd kN/sq.m	qb kN/sq.m	p m
-41	266	2.00	8.05	16.343

A sq.m	qD kN/sq.m	ft N/sq.mm	fb N/sq.mm	Vol. cu.m
206	2.541	0.031	0.413	22.1

s m	Nt kN/m	Mf kNm/m	Nf kN/m	Qf kN/m
0.00	36.2	0.23	-6.33	-1.36
0.45	19.8	0.62	-7.92	-0.45
0.90	7.0	0.69	-8.92	0.07
1.35	-1.8	0.60	-9.43	0.31
1.80	-7.0	0.44	-9.58	0.36
2.25	-9.6	0.28	-9.50	0.32
2.70	-10.4	0.15	-9.30	0.24
3.15	-10.3	0.06	-9.07	0.16
3.60	-9.8	0.01	-8.84	0.09
4.05	-9.2	-0.02	-8.64	0.04
4.95	-8.4	-0.03	-8.39	-0.01
5.85	-8.1	-0.02	-8.29	-0.02

TABLE 11 B

DIA.		RISE		RADIUS		ALPHA
m		m		m		deg.
16.100		2.007		17.145		28.00

h1	h2	S	b	d
mm	mm	m	mm	mm
75	180	2.957	250	700

y	db	qd	qb	p
mm	mm	kN/sq.m	kN/sq.m	m
-41	266	2.00	7.65	16.759

A	qD	ft	fb	Vol.
sq.m	kN/sq.m	N/sq.mm	N/sq.mm	cu.m
216	2.533	0.036	0.430	23.2

s	Nt	Mf	Nf	Qf
m	kN/m	kNm/m	kN/m	kN/m
0.00	37.9	0.24	-6.48	-1.40
0.45	21.0	0.65	-8.11	-0.47
0.90	7.7	0.72	-9.14	0.07
1.35	-1.5	0.63	-9.67	0.31
1.80	-7.0	0.47	-9.83	0.38
2.25	-9.8	0.30	-9.76	0.34
2.70	-10.7	0.17	-9.56	0.26
3.15	-10.7	0.07	-9.32	0.17
3.60	-10.2	0.01	-9.09	0.10
4.05	-9.5	-0.02	-8.89	0.05
4.95	-8.6	-0.03	-8.62	-0.01
5.85	-8.3	-0.02	-8.51	-0.02

TABLE 12 B

DIA.	RISE	RADIUS	ALPHA
m	m	m	deg.
16.600	2.070	17.677	28.00

h1	h2	S	b	d
mm	mm	m	mm	mm
75	180	3.003	250	700

y	db	qd	qb	p
mm	mm	kN/sq.m	kN/sq.m	m
-41	266	2.00	7.20	17.280

A	qD	ft	fb	Vol.
sq.m	kN/sq.m	N/sq.mm	N/sq.mm	cu.m
230	2.523	0.042	0.452	24.6

s	Nt	Mf	Nf	Qf
m	kN/m	kNm/m	kN/m	kN/m
0.00	40.1	0.26	-6.67	-1.45
0.45	22.5	0.68	-8.34	-0.49
0.90	8.6	0.77	-9.41	0.06
1.35	-1.1	0.67	-9.96	0.32
1.80	-7.0	0.50	-10.14	0.39
2.25	-10.0	0.33	-10.08	0.36
2.70	-11.1	0.19	-9.88	0.28
3.15	-11.1	0.09	-9.64	0.19
3.60	-10.6	0.02	-9.40	0.11
4.05	-10.0	-0.02	-9.19	0.05
4.95	-8.9	-0.03	-8.91	-0.01
6.30	-8.5	-0.01	-8.77	-0.01

TABLE 13 B

DIA.	RISE		RADIUS	ALPHA
m	m		m	deg.
17.100	2.132		18.210	28.00
h1	h2	S	b	d
mm	mm	m	mm	mm
75	180	3.047	250	700
y	db	qd	qb	p
mm	mm	kN/sq.m	kN/sq.m	m
-41	266	1.99	6.79	17.800
A	qD	ft	fb	Vol.
sq.m	kN/sq.m	N/sq.mm	N/sq.mm	cu.m
244	2.513	0.048	0.475	26.0
s	Nt	Mf	Nf	Qf
m	kN/m	kNm/m	kN/m	kN/m
0.00	42.4	0.28	-6.85	-1.50
0.45	24.1	0.72	-8.57	-0.52
0.90	9.5	0.81	-9.67	0.05
1.35	-0.6	0.71	-10.25	0.33
1.80	-6.9	0.54	-10.45	0.41
2.25	-10.2	0.36	-10.40	0.38
2.70	-11.5	0.21	-10.21	0.29
3.15	-11.6	0.10	-9.96	0.20
3.60	-11.0	0.03	-9.71	0.12
4.05	-10.4	-0.01	-9.49	0.06
5.40	-9.0	-0.03	-9.11	-0.01
6.30	-8.8	-0.02	-9.04	-0.02

TABLE 14 B

DIA.	RISE	RADIUS	ALPHA
m	m	m	deg.
17.600	2.194	18.742	28.00

h1	h2	S	b	d
mm	mm	m	mm	mm
75	180	3.092	250	700

y	db	qd	qb	p
mm	mm	kN/sq.m	kN/sq.m	m
-41	266	1.99	6.41	18.321

A	qD	ft	fb	Vol.
sq.m	kN/sq.m	N/sq.mm	N/sq.mm	cu.m
258	2.504	0.055	0.497	27.4

s	Nt	Mf	Nf	Qf
m	kN/m	kNm/m	kN/m	kN/m
0.00	44.8	0.30	-7.04	-1.55
0.45	25.7	0.75	-8.80	-0.55
0.90	10.6	0.85	-9.94	0.04
1.35	-0.2	0.76	-10.55	0.33
1.80	-6.8	0.58	-10.76	0.42
2.25	-10.4	0.40	-10.72	0.39
2.70	-11.8	0.23	-10.53	0.31
3.15	-12.0	0.11	-10.28	0.22
3.60	-11.5	0.03	-10.03	0.14
4.05	-10.8	-0.01	-9.80	0.07
4.50	-10.1	-0.03	-9.62	0.03
5.40	-9.2	-0.03	-9.40	-0.01
6.75	-9.0	-0.01	-9.30	-0.01

TABLE 15 B

DIA.	RISE	RADIUS	ALPHA
m	m	m	deg.
18.000	2.244	19.168	28.00

h1	h2	S	b	d
mm	mm	m	mm	mm
90	200	3.334	280	800

y	db	qd	qb	p
mm	mm	kN/sq.m	kN/sq.m	m
-6	343	2.37	8.82	18.737

A	qD	ft	fb	Vol.
sq.m	kN/sq.m	N/sq.mm	N/sq.mm	cu.m
270	2.929	0.056	0.403	33.6

s	Nt	Mf	Nf	Qf
m	kN/m	kNm/m	kN/m	kN/m
0.00	38.2	0.13	-7.28	-1.54
0.45	23.3	0.61	-8.80	-0.66
0.90	10.8	0.77	-9.85	-0.10
1.35	1.4	0.74	-10.48	0.21
1.80	-5.0	0.61	-10.78	0.34
2.25	-8.8	0.45	-10.84	0.36
2.70	-10.8	0.30	-10.74	0.31
3.15	-11.5	0.17	-10.56	0.24
3.60	-11.5	0.08	-10.34	0.17
4.05	-11.1	0.02	-10.13	0.10
4.50	-10.6	-0.01	-9.95	0.05
5.85	-9.5	-0.03	-9.60	-0.01
6.75	-9.3	-0.02	-9.52	-0.02

127

```
TABLE  16 B
DIA.           RISE        RADIUS         ALPHA
  m             m            m            deg.
18.500         2.307       19.701         28.00
  h1      h2       S          b            d
  mm      mm       m          mm           mm
  90      200      3.380      280          800
  y       db       qd         qb           p
  mm      mm       kN/sq.m    kN/sq.m      m
  -6      343      2.36       8.35         19.258
  A       qD       ft         fb           Vol.
  sq.m    kN/sq.m  N/sq.mm    N/sq.mm      cu.m
  286     2.920    0.061      0.421        35.3
  s       Nt       Mf         Nf           Qf
  m       kN/m     kNm/m      kN/m         kN/m
0.00     40.3     0.14       -7.47        -1.59
0.45     24.8     0.64       -9.03        -0.69
0.90     11.7     0.81       -10.11       -0.11
1.35      1.9     0.78       -10.77        0.21
1.80     -4.8     0.65       -11.08        0.35
2.25     -8.9     0.48       -11.15        0.37
2.70    -11.1     0.32       -11.05        0.33
3.15    -11.9     0.19       -10.87        0.26
3.60    -11.9     0.09       -10.66        0.18
4.05    -11.5     0.03       -10.44        0.11
4.50    -11.0    -0.01       -10.25        0.06
5.85     -9.8    -0.03        -9.89       -0.01
7.20     -9.5    -0.01        -9.78       -0.01
```

TABLE 17 B

DIA.	RISE	RADIUS	ALPHA
m	m	m	deg.
19.100	2.381	20.340	28.00

h1	h2	S	b	d
mm	mm	m	mm	mm
90	200	3.435	280	800

y	db	qd	qb	p
mm	mm	kN/sq.m	kN/sq.m	m
-6	343	2.36	7.83	19.882

A	qD	ft	fb	Vol.
sq.m	kN/sq.m	N/sq.mm	N/sq.mm	cu.m
304	2.909	0.068	0.444	37.5

s	Nt	Mf	Nf	Qf
m	kN/m	kNm/m	kN/m	kN/m
0.00	42.9	0.16	-7.69	-1.65
0.45	26.7	0.68	-9.30	-0.73
0.90	13.0	0.86	-10.42	-0.13
1.35	2.6	0.83	-11.11	0.21
1.80	-4.6	0.70	-11.44	0.36
2.25	-9.0	0.53	-11.52	0.39
2.70	-11.4	0.36	-11.44	0.35
3.15	-12.3	0.22	-11.25	0.28
3.60	-12.4	0.11	-11.04	0.20
4.05	-12.0	0.04	-10.81	0.13
4.50	-11.4	-0.01	-10.62	0.07
4.95	-10.9	-0.03	-10.45	0.03
5.85	-10.1	-0.04	-10.23	-0.01
7.20	-9.8	-0.02	-10.11	-0.02

```
TABLE  18  B
DIA.          RISE         RADIUS        ALPHA
  m             m             m           deg.
19.700        2.456        20.979        28.00
 h1         h2        S         b          d
 mm         mm        m         mm         mm
 90         200      3.488      280        800
  y         db        qd        qb          p
 mm         mm      kN/sq.m   kN/sq.m       m
 -6         343      2.36       7.36      20.507
  A         qD        ft        fb         Vol.
sq.m      kN/sq.m   N/sq.mm   N/sq.mm      cu.m
 324       2.898     0.076     0.468       39.8
  s         Nt        Mf        Nf          Qf
  m        kN/m      kNm/m     kN/m        kN/m
0.00       45.5      0.17      -7.91       -1.72
0.45       28.6      0.72      -9.57       -0.76
0.90       14.3      0.91     -10.73       -0.14
1.35        3.3      0.89     -11.44        0.21
1.80       -4.3      0.75     -11.80        0.37
2.25       -9.0      0.57     -11.89        0.41
2.70      -11.6      0.39     -11.82        0.37
3.15      -12.7      0.24     -11.64        0.30
3.60      -12.9      0.13     -11.41        0.22
4.05      -12.5      0.05     -11.19        0.14
4.50      -11.9      0.00     -10.98        0.08
4.95      -11.3     -0.03     -10.81        0.04
6.30      -10.2     -0.03     -10.50       -0.02
7.65      -10.1     -0.01     -10.42       -0.01
```

TABLE 19 B

DIA.		RISE		RADIUS		ALPHA
m		m		m		deg.
20.300		2.531		21.617		28.00

h1	h2	S	b	d
mm	mm	m	mm	mm
90	200	3.541	280	800

y	db	qd	qb	p
mm	mm	kN/sq.m	kN/sq.m	m
-6	343	2.35	6.93	21.131

A	qD	ft	fb	Vol.
sq.m	kN/sq.m	N/sq.mm	N/sq.mm	cu.m
344	2.888	0.084	0.492	42.1

s	Nt	Mf	Nf	Qf
m	kN/m	kNm/m	kN/m	kN/m
0.00	48.2	0.19	-8.14	-1.78
0.45	30.6	0.76	-9.84	-0.80
0.90	15.6	0.96	-11.03	-0.16
1.35	4.1	0.94	-11.78	0.21
1.80	-3.9	0.80	-12.16	0.38
2.25	-9.1	0.62	-12.27	0.42
2.70	-11.9	0.43	-12.20	0.39
3.15	-13.1	0.27	-12.02	0.32
3.60	-13.3	0.15	-11.79	0.23
4.05	-13.0	0.06	-11.56	0.16
4.50	-12.4	0.00	-11.35	0.09
4.95	-11.8	-0.03	-11.17	0.05
6.30	-10.6	-0.04	-10.84	-0.01
7.65	-10.4	-0.01	-10.75	-0.01

TABLE 20 B

DIA.		RISE		RADIUS		ALPHA
m		m		m		deg.
20.900		2.606		22.256		28.00
h1	h2		S		b	d
mm	mm		m		mm	mm
90	200		3.593		280	800
y	db	qd		qb		p
mm	mm	kN/sq.m		kN/sq.m		m
-6	343	2.35		6.54		21.756
A	qD	ft		fb		Vol.
sq.m	kN/sq.m	N/sq.mm		N/sq.mm		cu.m
364	2.878	0.092		0.516		44.4

s	Nt	Mf	Nf	Qf
m	kN/m	kNm/m	kN/m	kN/m
0.00	51.0	0.21	-8.36	-1.84
0.45	32.6	0.80	-10.10	-0.84
0.90	17.0	1.01	-11.34	-0.18
1.35	5.0	1.00	-12.11	0.21
1.80	-3.6	0.86	-12.52	0.39
2.25	-9.0	0.66	-12.64	0.44
2.70	-12.1	0.47	-12.58	0.41
3.15	-13.5	0.30	-12.40	0.34
3.60	-13.8	0.17	-12.18	0.25
4.05	-13.5	0.07	-11.94	0.17
4.50	-12.9	0.01	-11.72	0.11
4.95	-12.3	-0.02	-11.53	0.05
5.40	-11.7	-0.04	-11.38	0.02
6.75	-10.7	-0.03	-11.12	-0.02
8.10	-10.7	-0.01	-11.07	-0.01

TABLE 21 B

DIA.	RISE	RADIUS	ALPHA
m	m	m	deg.
21.500	2.681	22.895	28.00

h1	h2	S	b	d
mm	mm	m	mm	mm
90	200	3.644	280	800

y	db	qd	qb	p
mm	mm	kN/sq.m	kN/sq.m	m
-6	343	2.35	6.18	22.380

A	qD	ft	fb	Vol.
sq.m	kN/sq.m	N/sq.mm	N/sq.mm	cu.m
386	2.869	0.100	0.541	46.9

s	Nt	Mf	Nf	Qf
m	kN/m	kNm/m	kN/m	kN/m
0.00	53.8	0.22	-8.58	-1.91
0.45	34.7	0.84	-10.37	-0.88
0.90	18.5	1.07	-11.64	-0.20
1.35	5.8	1.06	-12.45	0.21
1.80	-3.2	0.91	-12.87	0.40
2.25	-9.0	0.71	-13.01	0.46
2.70	-12.4	0.51	-12.96	0.43
3.15	-13.9	0.33	-12.79	0.36
3.60	-14.3	0.19	-12.56	0.27
4.05	-14.0	0.09	-12.32	0.19
4.50	-13.4	0.02	-12.09	0.12
4.95	-12.7	-0.02	-11.89	0.06
5.40	-12.1	-0.04	-11.73	0.03
6.75	-11.1	-0.04	-11.46	-0.02
8.10	-11.0	-0.01	-11.40	-0.01

TABLE 22 B

DIA.	RISE		RADIUS	ALPHA
m	m		m	deg.
22.100	2.755		23.534	28.00
h1	h2	S	b	d
mm	mm	m	mm	mm
90	200	3.695	280	800
y	db	qd	qb	p
mm	mm	kN/sq.m	kN/sq.m	m
-6	343	2.34	5.85	23.005
A	qD	ft	fb	Vol.
sq.m	kN/sq.m	N/sq.mm	N/sq.mm	cu.m
408	2.860	0.109	0.566	49.4
s	Nt	Mf	Nf	Qf
m	kN/m	kNm/m	kN/m	kN/m
0.00	56.7	0.24	-8.79	-1.97
0.45	36.9	0.88	-10.63	-0.92
0.90	20.0	1.12	-11.94	-0.21
1.35	6.8	1.11	-12.78	0.20
1.80	-2.7	0.97	-13.23	0.41
2.25	-8.9	0.76	-13.38	0.48
2.70	-12.6	0.55	-13.34	0.45
3.15	-14.3	0.36	-13.17	0.38
3.60	-14.8	0.21	-12.94	0.29
4.05	-14.6	0.10	-12.69	0.21
4.50	-14.0	0.03	-12.46	0.13
4.95	-13.2	-0.02	-12.25	0.07
5.40	-12.6	-0.04	-12.08	0.03
6.75	-11.4	-0.04	-11.80	-0.02
8.55	-11.3	-0.01	-11.72	-0.01

TABLE 23 B

DIA.	RISE	RADIUS	ALPHA
m	m	m	deg.
22.800	2.843	24.280	28.00

h1	h2	S	b	d
mm	mm	m	mm	mm
90	200	3.753	280	800

y	db	qd	qb	p
mm	mm	kN/sq.m	kN/sq.m	m
-6	343	2.34	5.50	23.734

A	qD	ft	fb	Vol.
sq.m	kN/sq.m	N/sq.mm	N/sq.mm	cu.m
434	2.850	0.120	0.596	52.4

s	Nt	Mf	Nf	Qf
m	kN/m	kNm/m	kN/m	kN/m
0.00	60.1	0.26	-9.05	-2.05
0.45	39.5	0.93	-10.94	-0.97
0.90	21.8	1.19	-12.29	-0.24
1.35	7.9	1.18	-13.16	0.20
1.80	-2.2	1.04	-13.64	0.42
2.25	-8.8	0.83	-13.82	0.50
2.70	-12.8	0.60	-13.78	0.48
3.15	-14.7	0.40	-13.62	0.41
3.60	-15.4	0.24	-13.39	0.32
4.05	-15.2	0.12	-13.14	0.23
4.50	-14.6	0.04	-12.89	0.15
4.95	-13.8	-0.01	-12.68	0.08
5.40	-13.1	-0.04	-12.50	0.04
5.85	-12.5	-0.05	-12.36	0.01
7.20	-11.6	-0.04	-12.14	-0.02
8.55	-11.6	-0.01	-12.10	-0.01

```
TABLE  24  B
  DIA.        RISE         RADIUS        ALPHA
   m           m             m           deg.
 23.500      2.930        25.025        28.00
   h1         h2        S         b          d
   mm         mm        m         mm         mm
   90        200      3.810      280        800
   y          db        qd         qb         p
   mm         mm     kN/sq.m    kN/sq.m       m
   -6        343      2.34       5.17      24.462
   A          qD        ft         fb        Vol.
 sq.m      kN/sq.m   N/sq.mm    N/sq.mm     cu.m
  461       2.840     0.131      0.626      55.4
   s          Nt        Mf         Nf         Qf
   m         kN/m      kNm/m      kN/m       kN/m
 0.00       63.7       0.28      -9.30      -2.12
 0.45       42.2       0.98     -11.24      -1.01
 0.90       23.7       1.25     -12.64      -0.26
 1.35        9.1       1.26     -13.55       0.20
 1.80       -1.5       1.11     -14.05       0.43
 2.25       -8.7       0.89     -14.25       0.52
 2.70      -13.0       0.66     -14.23       0.50
 3.15      -15.2       0.45     -14.07       0.43
 3.60      -15.9       0.27     -13.84       0.34
 4.05      -15.8       0.14     -13.58       0.25
 4.50      -15.2       0.05     -13.33       0.16
 4.95      -14.4      -0.01     -13.11       0.10
 5.40      -13.7      -0.04     -12.92       0.05
 5.85      -13.0      -0.05     -12.77       0.01
 7.20      -12.0      -0.04     -12.53      -0.02
 9.00      -12.0      -0.01     -12.48      -0.01
```

TABLE 25 B

DIA.	RISE	RADIUS	ALPHA
m	m	m	deg.
24.000	2.992	25.558	28.00

h1	h2	S	b	d
mm	mm	m	mm	mm
100	250	4.230	300	1000

y	db	qd	qb	p
mm	mm	kN/sq.m	kN/sq.m	m
43	487	2.68	6.12	24.983

A	qD	ft	fb	Vol.
sq.m	kN/sq.m	N/sq.mm	N/sq.mm	cu.m
481	3.411	0.105	0.465	69.5

s	Nt	Mf	Nf	Qf
m	kN/m	kNm/m	kN/m	kN/m
0.00	54.9	-0.10	-9.27	-2.29
0.45	39.1	0.69	-10.96	-1.30
0.90	24.7	1.11	-12.28	-0.57
1.35	12.6	1.25	-13.23	-0.08
1.80	2.9	1.21	-13.86	0.23
2.25	-4.3	1.06	-14.22	0.39
2.70	-9.3	0.87	-14.37	0.45
3.15	-12.5	0.67	-14.36	0.44
3.60	-14.3	0.48	-14.25	0.39
4.05	-15.2	0.32	-14.08	0.32
4.50	-15.3	0.19	-13.87	0.25
4.95	-15.1	0.09	-13.66	0.18
5.40	-14.6	0.03	-13.46	0.12
5.85	-14.1	-0.02	-13.27	0.07
7.65	-12.6	-0.05	-12.82	-0.01
9.00	-12.4	-0.02	-12.71	-0.02

```
TABLE  26  B
  DIA.         RISE         RADIUS       ALPHA
   m            m             m          deg.
24.700       3.080        26.303        28.00
  h1          h2           S            b          d
  mm          mm           m            mm         mm
 100         250          4.291        300        1000
   y          db           qd           qb         p
  mm          mm        kN/sq.m      kN/sq.m       m
  43         487          2.67         5.78      25.712
   A          qD           ft           fb        Vol.
 sq.m      kN/sq.m      N/sq.mm      N/sq.mm      cu.m
 509        3.398        0.113        0.488       73.3
   s          Nt           Mf           Nf         Qf
   m         kN/m        kNm/m         kN/m       kN/m
 0.00        58.0        -0.09        -9.52      -2.37
 0.45        41.6         0.73       -11.26      -1.35
 0.90        26.6         1.17       -12.61      -0.61
 1.35        13.9         1.32       -13.60      -0.10
 1.80         3.8         1.28       -14.25       0.23
 2.25        -3.9         1.13       -14.64       0.40
 2.70        -9.2         0.94       -14.80       0.47
 3.15       -12.7         0.73       -14.80       0.46
 3.60       -14.7         0.53       -14.70       0.41
 4.05       -15.6         0.36       -14.52       0.35
 4.50       -15.9         0.22       -14.31       0.27
 4.95       -15.6         0.11       -14.09       0.20
 5.40       -15.2         0.04       -13.89       0.14
 5.85       -14.6        -0.01       -13.70       0.08
 6.30       -14.1        -0.04       -13.54       0.05
 7.65       -13.0        -0.05       -13.22      -0.01
 9.45       -12.7        -0.02       -13.08      -0.02
```

TABLE 27 B

DIA.	RISE	RADIUS	ALPHA
m	m	m	deg.
25.500	3.179	27.155	28.00

h1	h2	S	b	d
mm	mm	m	mm	mm
100	250	4.360	300	1000

y	db	qd	qb	p
mm	mm	kN/sq.m	kN/sq.m	m
43	487	2.67	5.42	26.544

A	qD	ft	fb	Vol.
sq.m	kN/sq.m	N/sq.mm	N/sq.mm	cu.m
543	3.383	0.123	0.515	77.8

s	Nt	Mf	Nf	Qf
m	kN/m	kNm/m	kN/m	kN/m
0.00	61.7	-0.08	-9.80	-2.46
0.45	44.6	0.78	-11.59	-1.42
0.90	28.9	1.24	-12.99	-0.65
1.35	15.5	1.40	-14.01	-0.11
1.80	4.8	1.37	-14.70	0.22
2.25	-3.3	1.22	-15.11	0.41
2.70	-9.1	1.01	-15.29	0.49
3.15	-12.8	0.79	-15.30	0.49
3.60	-15.1	0.58	-15.20	0.44
4.05	-16.2	0.40	-15.03	0.37
4.50	-16.5	0.25	-14.82	0.29
4.95	-16.3	0.14	-14.59	0.22
5.40	-15.8	0.05	-14.38	0.15
5.85	-15.3	0.00	-14.18	0.10
6.30	-14.7	-0.04	-14.01	0.05
8.10	-13.2	-0.05	-13.61	-0.02
9.90	-13.1	-0.02	-13.51	-0.02

TABLE 28 B

DIA.		RISE		RADIUS		ALPHA
m		m		m		deg.
26.200		3.267		27.900		28.00

h1	h2	S	b	d
mm	mm	m	mm	mm
100	250	4.419	300	1000

y	db	qd	qb	p
mm	mm	kN/sq.m	kN/sq.m	m
43	487	2.66	5.14	27.273

A	qD	ft	fb	Vol.
sq.m	kN/sq.m	N/sq.mm	N/sq.mm	cu.m
573	3.371	0.132	0.539	81.8

s	Nt	Mf	Nf	Qf
m	kN/m	kNm/m	kN/m	kN/m
0.00	65.1	-0.07	-10.05	-2.54
0.45	47.2	0.82	-11.88	-1.47
0.90	30.9	1.30	-13.32	-0.68
1.35	16.9	1.47	-14.38	-0.13
1.80	5.7	1.44	-15.09	0.22
2.25	-2.8	1.29	-15.52	0.42
2.70	-8.9	1.08	-15.71	0.50
3.15	-13.0	0.85	-15.74	0.51
3.60	-15.4	0.63	-15.64	0.46
4.05	-16.6	0.44	-15.47	0.39
4.50	-17.0	0.28	-15.26	0.32
4.95	-16.9	0.16	-15.03	0.24
5.40	-16.4	0.07	-14.81	0.17
5.85	-15.8	0.00	-14.61	0.11
6.30	-15.2	-0.04	-14.43	0.06
6.75	-14.7	-0.06	-14.28	0.03
8.10	-13.6	-0.06	-14.00	-0.02
9.90	-13.4	-0.02	-13.89	-0.02

TABLE 29 B

DIA. m		RISE m	RADIUS m	ALPHA deg.
27.000		3.366	28.752	28.00

h1 mm	h2 mm	S m	b mm	d mm
100	250	4.486	300	1000

y mm	db mm	qd kN/sq.m	qb kN/sq.m	p m
43	487	2.66	4.84	28.106

A sq.m	qD kN/sq.m	ft N/sq.mm	fb N/sq.mm	Vol. cu.m
608	3.358	0.143	0.567	86.5

s m	Nt kN/m	Mf kNm/m	Nf kN/m	Qf kN/m
0.00	68.9	-0.05	-10.33	-2.63
0.45	50.3	0.87	-12.21	-1.54
0.90	33.3	1.37	-13.69	-0.72
1.35	18.7	1.56	-14.79	-0.15
1.80	6.9	1.53	-15.53	0.22
2.25	-2.2	1.38	-15.98	0.43
2.70	-8.7	1.17	-16.20	0.52
3.15	-13.1	0.93	-16.24	0.53
3.60	-15.7	0.70	-16.15	0.49
4.05	-17.1	0.49	-15.98	0.42
4.50	-17.6	0.32	-15.77	0.34
4.95	-17.5	0.18	-15.54	0.26
5.40	-17.1	0.08	-15.31	0.19
5.85	-16.5	0.01	-15.10	0.12
6.30	-15.9	-0.03	-14.91	0.08
6.75	-15.3	-0.06	-14.75	0.04
8.55	-13.9	-0.05	-14.39	-0.02
10.35	-13.9	-0.02	-14.32	-0.02

TABLE 30 B

DIA.		RISE		RADIUS		ALPHA
m		m		m		deg.
27.800		3.466		29.604		28.00

h1	h2	S	b	d
mm	mm	m	mm	mm
100	250	4.552	300	1000

y	db	qd	qb	p
mm	mm	kN/sq.m	kN/sq.m	m
43	487	2.66	4.56	28.938

A	qD	ft	fb	Vol.
sq.m	kN/sq.m	N/sq.mm	N/sq.mm	cu.m
645	3.345	0.154	0.596	91.4

s	Nt	Mf	Nf	Qf
m	kN/m	kNm/m	kN/m	kN/m
0.00	72.9	-0.04	-10.60	-2.72
0.45	53.5	0.92	-12.54	-1.60
0.90	35.8	1.44	-14.06	-0.76
1.35	20.5	1.64	-15.20	-0.17
1.80	8.1	1.63	-15.97	0.22
2.25	-1.5	1.47	-16.45	0.44
2.70	-8.4	1.25	-16.69	0.54
3.15	-13.1	1.00	-16.74	0.56
3.60	-16.1	0.76	-16.66	0.52
4.05	-17.6	0.54	-16.49	0.45
4.50	-18.3	0.36	-16.28	0.37
4.95	-18.2	0.21	-16.04	0.28
5.40	-17.8	0.10	-15.81	0.21
5.85	-17.2	0.02	-15.59	0.14
6.30	-16.5	-0.03	-15.39	0.09
6.75	-15.9	-0.06	-15.23	0.05
7.20	-15.3	-0.07	-15.09	0.02
9.00	-14.2	-0.05	-14.80	-0.02
10.80	-14.3	-0.01	-14.75	-0.01

TABLE 31 B

DIA.	RISE	RADIUS	ALPHA
m	m	m	deg.
28.700	3.578	30.563	28.00

h1	h2	S	b	d
mm	mm	m	mm	mm
100	250	4.625	300	1000

y	db	qd	qb	p
mm	mm	kN/sq.m	kN/sq.m	m
43	487	2.65	4.28	29.875

A	qD	ft	fb	Vol.
sq.m	kN/sq.m	N/sq.mm	N/sq.mm	cu.m
687	3.331	0.167	0.628	97.0

s	Nt	Mf	Nf	Qf
m	kN/m	kNm/m	kN/m	kN/m
0.00	77.5	-0.02	-10.91	-2.83
0.45	57.3	0.98	-12.91	-1.68
0.90	38.6	1.53	-14.48	-0.81
1.35	22.6	1.74	-15.66	-0.19
1.80	9.5	1.73	-16.47	0.21
2.25	-0.6	1.58	-16.97	0.45
2.70	-8.1	1.35	-17.23	0.56
3.15	-13.2	1.09	-17.30	0.58
3.60	-16.4	0.84	-17.23	0.55
4.05	-18.2	0.60	-17.06	0.48
4.50	-18.9	0.41	-16.85	0.40
4.95	-19.0	0.25	-16.61	0.31
5.40	-18.6	0.13	-16.37	0.23
5.85	-17.9	0.04	-16.14	0.16
6.30	-17.2	-0.02	-15.94	0.10
6.75	-16.6	-0.05	-15.76	0.06
7.20	-15.9	-0.07	-15.62	0.02
9.00	-14.7	-0.05	-15.30	-0.02
10.80	-14.7	-0.01	-15.24	-0.02

```
TABLE  32  B
```

DIA.	RISE	RADIUS	ALPHA
m	m	m	deg.
29.500	3.678	31.415	28.00

h1	h2	S	b	d
mm	mm	m	mm	mm
100	250	4.689	300	1000

y	db	qd	qb	p
mm	mm	kN/sq.m	kN/sq.m	m
43	487	2.65	4.05	30.708

A	qD	ft	fb	Vol.
sq.m	kN/sq.m	N/sq.mm	N/sq.mm	cu.m
726	3.319	0.178	0.658	102.1

s	Nt	Mf	Nf	Qf
m	kN/m	kNm/m	kN/m	kN/m
0.00	81.7	-0.01	-11.19	-2.92
0.45	60.7	1.03	-13.23	-1.74
0.90	41.3	1.60	-14.85	-0.85
1.35	24.6	1.84	-16.06	-0.22
1.80	10.9	1.83	-16.91	0.20
2.25	0.2	1.68	-17.44	0.45
2.70	-7.7	1.44	-17.71	0.58
3.15	-13.2	1.17	-17.79	0.61
3.60	-16.7	0.91	-17.73	0.57
4.05	-18.7	0.66	-17.57	0.51
4.50	-19.5	0.45	-17.36	0.42
4.95	-19.7	0.28	-17.12	0.33
5.40	-19.3	0.15	-16.87	0.25
5.85	-18.6	0.05	-16.64	0.18
6.30	-17.9	-0.01	-16.43	0.12
6.75	-17.2	-0.05	-16.24	0.07
7.20	-16.5	-0.07	-16.09	0.03
7.65	-16.0	-0.08	-15.97	0.01
9.45	-15.0	-0.05	-15.71	-0.03
11.25	-15.1	-0.01	-15.67	-0.01

TABLE 33 B

DIA.	RISE	RADIUS	ALPHA
m	m	m	deg.
30.000	3.740	31.947	28.00

h1	h2	S	b	d
mm	mm	m	mm	mm
130	290	5.180	300	1200

y	db	qd	qb	p
mm	mm	kN/sq.m	kN/sq.m	m
115	642	3.40	6.62	31.229

A	qD	ft	fb	Vol.
sq.m	kN/sq.m	N/sq.mm	N/sq.mm	cu.m
751	4.169	0.157	0.528	132.7

s	Nt	Mf	Nf	Qf
m	kN/m	kNm/m	kN/m	kN/m
0.00	73.8	-0.64	-10.99	-3.18
0.45	57.6	0.53	-12.84	-2.09
0.90	41.9	1.27	-14.38	-1.23
1.35	27.7	1.67	-15.60	-0.57
1.80	15.5	1.82	-16.53	-0.10
2.25	5.4	1.78	-17.19	0.22
2.70	-2.6	1.64	-17.61	0.41
3.15	-8.7	1.42	-17.85	0.51
3.60	-13.0	1.18	-17.93	0.54
4.05	-16.0	0.94	-17.89	0.52
4.50	-17.8	0.72	-17.78	0.47
4.95	-18.8	0.52	-17.61	0.41
5.40	-19.1	0.35	-17.41	0.34
5.85	-19.0	0.21	-17.19	0.27
6.30	-18.7	0.11	-16.98	0.20
6.75	-18.2	0.03	-16.78	0.14
7.20	-17.6	-0.02	-16.60	0.09
7.65	-17.1	-0.05	-16.45	0.06
9.45	-15.7	-0.07	-16.03	-0.02
11.70	-15.5	-0.02	-15.89	-0.02

TABLE 34 B

DIA. m		RISE m	RADIUS m	ALPHA deg.
30.900		3.853	32.905	28.00

h1 mm	h2 mm	S m	b mm	d mm
130	290	5.257	300	1200

y mm	db mm	qd kN/sq.m	qb kN/sq.m	p m
115	642	3.39	6.24	32.165

A sq.m	qD kN/sq.m	ft N/sq.mm	fb N/sq.mm	Vol. cu.m
797	4.155	0.168	0.556	140.3

s m	Nt kN/m	Mf kNm/m	Nf kN/m	Qf kN/m
0.00	78.2	-0.64	-11.29	-3.29
0.45	61.2	0.58	-13.19	-2.17
0.90	44.9	1.35	-14.77	-1.29
1.35	30.0	1.77	-16.04	-0.61
1.80	17.2	1.92	-17.00	-0.12
2.25	6.6	1.90	-17.69	0.22
2.70	-1.9	1.75	-18.13	0.42
3.15	-8.3	1.53	-18.39	0.53
3.60	-13.0	1.28	-18.48	0.57
4.05	-16.2	1.03	-18.46	0.55
4.50	-18.2	0.79	-18.34	0.50
4.95	-19.3	0.58	-18.18	0.44
5.40	-19.8	0.40	-17.97	0.36
5.85	-19.7	0.25	-17.76	0.29
6.30	-19.4	0.13	-17.54	0.22
6.75	-18.9	0.05	-17.34	0.16
7.20	-18.4	-0.01	-17.15	0.11
7.65	-17.8	-0.05	-16.98	0.07
9.90	-16.0	-0.07	-16.48	-0.02
11.70	-15.9	-0.03	-16.38	-0.02

TABLE 35 B

DIA.	RISE	RADIUS	ALPHA
m	m	m	deg.
31.800	3.965	33.864	28.00

h1	h2	S	b	d
mm	mm	m	mm	mm
130	290	5.333	300	1200

y	db	qd	qb	p
mm	mm	kN/sq.m	kN/sq.m	m
115	642	3.39	5.89	33.102

A	qD	ft	fb	Vol.
sq.m	kN/sq.m	N/sq.mm	N/sq.mm	cu.m
844	4.141	0.179	0.584	148.0

s	Nt	Mf	Nf	Qf
m	kN/m	kNm/m	kN/m	kN/m
0.00	82.7	-0.64	-11.59	-3.40
0.45	65.0	0.62	-13.54	-2.26
0.90	47.9	1.42	-15.17	-1.35
1.35	32.4	1.86	-16.47	-0.65
1.80	19.0	2.03	-17.47	-0.14
2.25	7.8	2.01	-18.18	0.21
2.70	-1.1	1.86	-18.66	0.43
3.15	-8.0	1.64	-18.93	0.55
3.60	-13.0	1.38	-19.03	0.59
4.05	-16.4	1.11	-19.02	0.58
4.50	-18.7	0.86	-18.91	0.53
4.95	-19.9	0.64	-18.75	0.47
5.40	-20.4	0.45	-18.54	0.39
5.85	-20.5	0.29	-18.32	0.31
6.30	-20.2	0.16	-18.10	0.24
6.75	-19.7	0.07	-17.89	0.18
7.20	-19.1	0.00	-17.69	0.12
7.65	-18.5	-0.04	-17.52	0.08
8.10	-17.9	-0.07	-17.37	0.04
9.90	-16.5	-0.08	-16.99	-0.02
12.15	-16.3	-0.02	-16.86	-0.02

TABLE 36 B

DIA.	RISE	RADIUS	ALPHA
m	m	m	deg.
32.800	4.090	34.929	28.00

h1	h2	S	b	d
mm	mm	m	mm	mm
130	290	5.417	300	1200

y	db	qd	qb	p
mm	mm	kN/sq.m	kN/sq.m	m
115	642	3.39	5.54	34.143

A	qD	ft	fb	Vol.
sq.m	kN/sq.m	N/sq.mm	N/sq.mm	cu.m
898	4.126	0.193	0.616	157.0

s	Nt	Mf	Nf	Qf
m	kN/m	kNm/m	kN/m	kN/m
0.00	87.8	-0.64	-11.92	-3.52
0.45	69.3	0.67	-13.93	-2.35
0.90	51.5	1.51	-15.60	-1.41
1.35	35.2	1.97	-16.95	-0.69
1.80	21.1	2.16	-17.98	-0.16
2.25	9.3	2.14	-18.73	0.21
2.70	-0.2	1.99	-19.23	0.44
3.15	-7.5	1.76	-19.52	0.57
3.60	-12.9	1.49	-19.65	0.62
4.05	-16.7	1.22	-19.64	0.61
4.50	-19.1	0.95	-19.54	0.57
4.95	-20.5	0.71	-19.38	0.50
5.40	-21.2	0.50	-19.18	0.42
5.85	-21.3	0.33	-18.95	0.34
6.30	-21.0	0.19	-18.73	0.27
6.75	-20.5	0.09	-18.51	0.20
7.20	-19.9	0.01	-18.30	0.14
7.65	-19.3	-0.04	-18.12	0.09
8.10	-18.7	-0.07	-17.96	0.05
10.35	-16.9	-0.07	-17.50	-0.03
12.60	-16.8	-0.02	-17.40	-0.02

TABLE 37 B

DIA.	RISE	RADIUS	ALPHA
m	m	m	deg.
33.800	4.214	35.994	28.00

h1	h2	S	b	d
mm	mm	m	mm	mm
130	290	5.499	300	1200

y	db	qd	qb	p
mm	mm	kN/sq.m	kN/sq.m	m
115	642	3.38	5.22	35.184

A	qD	ft	fb	Vol.
sq.m	kN/sq.m	N/sq.mm	N/sq.mm	cu.m
953	4.112	0.206	0.648	166.1

s	Nt	Mf	Nf	Qf
m	kN/m	kNm/m	kN/m	kN/m
0.00	93.0	-0.64	-12.25	-3.65
0.45	73.7	0.72	-14.31	-2.45
0.90	55.1	1.60	-16.04	-1.48
1.35	38.1	2.09	-17.43	-0.74
1.80	23.3	2.29	-18.50	-0.18
2.25	10.9	2.28	-19.28	0.20
2.70	0.9	2.13	-19.81	0.45
3.15	-6.9	1.89	-20.12	0.59
3.60	-12.7	1.61	-20.26	0.64
4.05	-16.8	1.32	-20.27	0.64
4.50	-19.5	1.04	-20.18	0.60
4.95	-21.1	0.79	-20.01	0.53
5.40	-21.9	0.57	-19.81	0.45
5.85	-22.1	0.38	-19.58	0.37
6.30	-21.8	0.23	-19.35	0.29
6.75	-21.3	0.11	-19.12	0.22
7.20	-20.7	0.03	-18.91	0.16
7.65	-20.1	-0.03	-18.72	0.11
8.10	-19.4	-0.07	-18.55	0.06
8.55	-18.8	-0.09	-18.41	0.03
10.80	-17.3	-0.07	-18.01	-0.03
13.05	-17.4	-0.02	-17.94	-0.02

TABLE 38 B

DIA.		RISE		RADIUS		ALPHA
m		m		m		deg.
34.800		4.339		37.059		28.00

h1	h2	S		b	d
mm	mm	m		mm	mm
130	290	5.579		300	1200

y	db	qd		qb	p
mm	mm	kN/sq.m		kN/sq.m	m
115	642	3.38		4.92	36.225

A	qD	ft	fb	Vol.
sq.m	kN/sq.m	N/sq.mm	N/sq.mm	cu.m
1010	4.099	0.220	0.682	175.5

s	Nt	Mf	Nf	Qf
m	kN/m	kNm/m	kN/m	kN/m
0.00	98.4	-0.64	-12.58	-3.78
0.45	78.2	0.77	-14.69	-2.55
0.90	58.8	1.68	-16.47	-1.55
1.35	41.1	2.20	-17.90	-0.78
1.80	25.6	2.42	-19.01	-0.21
2.25	12.6	2.41	-19.83	0.19
2.70	2.0	2.26	-20.38	0.45
3.15	-6.3	2.02	-20.72	0.60
3.60	-12.5	1.73	-20.87	0.67
4.05	-17.0	1.43	-20.89	0.67
4.50	-19.9	1.14	-20.81	0.63
4.95	-21.7	0.87	-20.65	0.57
5.40	-22.6	0.63	-20.45	0.49
5.85	-22.9	0.43	-20.22	0.40
6.30	-22.7	0.27	-19.98	0.32
6.75	-22.2	0.14	-19.75	0.25
7.20	-21.6	0.05	-19.53	0.18
7.65	-20.9	-0.02	-19.33	0.12
8.10	-20.2	-0.06	-19.15	0.08
8.55	-19.6	-0.09	-19.00	0.04
9.00	-19.0	-0.10	-18.87	0.01
11.25	-17.7	-0.07	-18.53	-0.03
13.50	-17.9	-0.01	-18.48	-0.01

TABLE 39 B

DIA.	RISE	RADIUS	ALPHA
m	m	m	deg.
35.000	4.364	37.272	28.00

h1	h2	S	b	d
mm	mm	m	mm	mm
130	290	5.595	300	1200

y	db	qd	qb	p
mm	mm	kN/sq.m	kN/sq.m	m
115	642	3.38	4.87	36.433

A	qD	ft	fb	Vol.
sq.m	kN/sq.m	N/sq.mm	N/sq.mm	cu.m
1022	4.096	0.223	0.688	177.4

s	Nt	Mf	Nf	Qf
m	kN/m	kNm/m	kN/m	kN/m
0.00	99.5	-0.64	-12.64	-3.80
0.45	79.2	0.78	-14.77	-2.57
0.90	59.6	1.70	-16.55	-1.57
1.35	41.7	2.22	-18.00	-0.79
1.80	26.1	2.44	-19.12	-0.21
2.25	12.9	2.44	-19.94	0.19
2.70	2.2	2.29	-20.50	0.45
3.15	-6.2	2.05	-20.84	0.61
3.60	-12.5	1.76	-21.00	0.67
4.05	-17.0	1.45	-21.02	0.68
4.50	-20.0	1.16	-20.93	0.64
4.95	-21.9	0.89	-20.78	0.57
5.40	-22.8	0.65	-20.57	0.49
5.85	-23.1	0.44	-20.34	0.41
6.30	-22.9	0.28	-20.10	0.33
6.75	-22.4	0.15	-19.87	0.25
7.20	-21.7	0.05	-19.65	0.18
7.65	-21.0	-0.02	-19.45	0.13
8.10	-20.3	-0.06	-19.27	0.08
8.55	-19.7	-0.09	-19.12	0.04
9.00	-19.2	-0.10	-18.99	0.02
11.25	-17.8	-0.07	-18.64	-0.03
13.50	-18.0	-0.01	-18.59	-0.02

TABLE 40 B

DIA.		RISE		RADIUS		ALPHA
m		m		m		deg.
36.100		4.501		38.443		28.00

h1	h2	S	b	d
mm	mm	m	mm	mm
130	290	5.683	300	1200

y	db	qd	qb	p
mm	mm	kN/sq.m	kN/sq.m	m
115	642	3.37	4.57	37.578

A	qD	ft	fb	Vol.
sq.m	kN/sq.m	N/sq.mm	N/sq.mm	cu.m
1087	4.082	0.240	0.726	188.1

s	Nt	Mf	Nf	Qf
m	kN/m	kNm/m	kN/m	kN/m
0.00	105.6	-0.64	-13.00	-3.94
0.45	84.3	0.84	-15.18	-2.67
0.90	63.9	1.80	-17.03	-1.64
1.35	45.2	2.35	-18.52	-0.84
1.80	28.7	2.59	-19.68	-0.24
2.25	14.9	2.60	-20.53	0.18
2.70	3.5	2.45	-21.12	0.46
3.15	-5.4	2.20	-21.49	0.62
3.60	-12.2	1.90	-21.67	0.70
4.05	-17.1	1.58	-21.70	0.71
4.50	-20.4	1.27	-21.63	0.67
4.95	-22.5	0.98	-21.48	0.61
5.40	-23.6	0.72	-21.27	0.53
5.85	-23.9	0.51	-21.04	0.44
6.30	-23.8	0.33	-20.80	0.36
6.75	-23.3	0.18	-20.56	0.28
7.20	-22.7	0.08	-20.33	0.21
7.65	-22.0	0.00	-20.12	0.15
8.10	-21.2	-0.06	-19.93	0.09
8.55	-20.5	-0.09	-19.77	0.05
9.00	-19.9	-0.11	-19.63	0.02
11.70	-18.3	-0.06	-19.22	-0.03
13.95	-18.5	-0.01	-19.19	-0.01

Tables 1C to 40C   For Type C Domes

TABLE 1 C

DIA.		RISE		RADIUS		ALPHA
m		m		m		deg.
12.000		1.496		12.779		28.00
h1	h2		S	b		d
mm	mm		m	mm		mm
75	180		2.553	250		700
y	db	qd		qb		p
mm	mm	kN/sq.m		kN/sq.m		m
-308	0	2.03		13.78		12.491
A	qD		ft	fb		Vol.
sq.m	kN/sq.m		N/sq.mm	N/sq.mm		cu.m
120	2.636		-0.170	0.319		13.4
s	Nt		Mf	Nf		Qf
m	kN/m		kNm/m	kN/m		kN/m
0.00	43.4		0.23	-3.41		-1.79
0.45	22.6		0.72	-5.74		-0.48
0.90	7.2		0.76	-7.12		0.20
1.35	-2.4		0.60	-7.74		0.45
1.80	-7.4		0.40	-7.83		0.45
2.25	-9.1		0.21	-7.62		0.35
2.70	-9.2		0.09	-7.30		0.22
3.60	-7.6		-0.02	-6.67		0.04
4.50	-6.5		-0.03	-6.30		-0.02

TABLE 2 C

DIA. m		RISE m	RADIUS m	ALPHA deg.
12.400		1.546	13.205	28.00

h1 mm	h2 mm	S m	b mm	d mm
75	180	2.595	250	700

y mm	db mm	qd kN/sq.m	qb kN/sq.m	p m
-308	0	2.03	12.90	12.908

A sq.m	qD kN/sq.m	ft N/sq.mm	fb N/sq.mm	Vol. cu.m
128	2.624	-0.175	0.337	14.3

s m	Nt kN/m	Mf kNm/m	Nf kN/m	Qf kN/m
0.00	45.8	0.27	-3.56	-1.84
0.45	24.1	0.77	-5.94	-0.49
0.90	7.9	0.81	-7.36	0.20
1.35	-2.2	0.65	-8.00	0.46
1.80	-7.5	0.43	-8.10	0.47
2.25	-9.5	0.24	-7.90	0.37
2.70	-9.6	0.10	-7.57	0.24
3.15	-8.9	0.02	-7.22	0.13
3.60	-8.0	-0.02	-6.92	0.05
4.50	-6.8	-0.03	-6.54	-0.01

TABLE 3 C

DIA.	RISE	RADIUS	ALPHA
m	m	m	deg.
12.700	1.583	13.524	28.00

h1	h2	S	b	d
mm	mm	m	mm	mm
75	180	2.626	250	700

y	db	qd	qb	p
mm	mm	kN/sq.m	kN/sq.m	m
-308	0	2.02	12.30	13.220

A	qD	ft	fb	Vol.
sq.m	kN/sq.m	N/sq.mm	N/sq.mm	cu.m
135	2.615	-0.179	0.350	14.9

s	Nt	Mf	Nf	Qf
m	kN/m	kNm/m	kN/m	kN/m
0.00	47.7	0.30	-3.67	-1.87
0.45	25.3	0.80	-6.09	-0.51
0.90	8.5	0.85	-7.54	0.21
1.35	-2.1	0.68	-8.19	0.48
1.80	-7.6	0.46	-8.31	0.49
2.25	-9.7	0.26	-8.11	0.39
2.70	-9.9	0.11	-7.77	0.26
3.15	-9.2	0.02	-7.42	0.14
4.05	-7.5	-0.04	-6.88	0.01
4.95	-6.7	-0.03	-6.61	-0.02

```
TABLE  4  C
```

DIA.		RISE		RADIUS		ALPHA
m		m		m		deg.
13.100		1.633		13.950		28.00

h1	h2	S	b	d
mm	mm	m	mm	mm
75	180	2.667	250	700

y	db	qd	qb	p
mm	mm	kN/sq.m	kN/sq.m	m
-308	0	2.02	11.56	13.636

A	qD	ft	fb	Vol.
sq.m	kN/sq.m	N/sq.mm	N/sq.mm	cu.m
143	2.604	-0.183	0.368	15.8

s	Nt	Mf	Nf	Qf
m	kN/m	kNm/m	kN/m	kN/m
0.00	50.2	0.34	-3.81	-1.91
0.45	26.8	0.85	-6.29	-0.52
0.90	9.3	0.90	-7.77	0.21
1.35	-1.8	0.73	-8.45	0.50
1.80	-7.7	0.49	-8.58	0.52
2.25	-10.1	0.28	-8.38	0.41
2.70	-10.4	0.13	-8.05	0.28
3.15	-9.7	0.03	-7.68	0.16
4.05	-7.9	-0.04	-7.12	0.02
4.95	-6.9	-0.03	-6.84	-0.02

TABLE 5 C

DIA. m	RISE m	RADIUS m	ALPHA deg.
13.500	1.683	14.376	28.00

h1 mm	h2 mm	S m	b mm	d mm
75	180	2.708	250	700

y mm	db mm	qd kN/sq.m	qb kN/sq.m	p m
-308	0	2.02	10.89	14.053

A sq.m	qD kN/sq.m	ft N/sq.mm	fb N/sq.mm	Vol. cu.m
152	2.594	-0.187	0.386	16.7

s m	Nt kN/m	Mf kNm/m	Nf kN/m	Qf kN/m
0.00	52.7	0.37	-3.96	-1.96
0.45	28.4	0.91	-6.49	-0.54
0.90	10.2	0.96	-8.01	0.22
1.35	-1.5	0.78	-8.71	0.52
1.80	-7.8	0.53	-8.85	0.54
2.25	-10.4	0.31	-8.66	0.44
2.70	-10.8	0.14	-8.32	0.30
3.15	-10.1	0.04	-7.95	0.17
4.05	-8.2	-0.04	-7.36	0.02
4.95	-7.2	-0.03	-7.07	-0.02

```
TABLE  6  C
  DIA.          RISE          RADIUS         ALPHA
   m             m              m            deg.
13.900         1.733         14.802         28.00
  h1          h2        S           b            d
  mm          mm        m           mm           mm
  75          180      2.748        250          700
   y          db        qd           qb          p
  mm          mm     kN/sq.m      kN/sq.m        m
 -308          0       2.01        10.27       14.469
   A          qD        ft           fb         Vol.
 sq.m      kN/sq.m   N/sq.mm      N/sq.mm       cu.m
  161        2.583    -0.191        0.404        17.6
   s          Nt        Mf           Nf          Qf
   m         kN/m      kNm/m        kN/m        kN/m
 0.00        55.2      0.41        -4.10        -2.00
 0.45        30.0      0.96        -6.68        -0.55
 0.90        11.0      1.01        -8.24         0.22
 1.35        -1.2      0.83        -8.97         0.53
 1.80        -7.9      0.57        -9.13         0.57
 2.25       -10.8      0.34        -8.94         0.46
 2.70       -11.2      0.16        -8.60         0.32
 3.15       -10.6      0.05        -8.22         0.19
 3.60        -9.5     -0.02        -7.88         0.09
 4.50        -7.9     -0.04        -7.42        -0.01
 5.40        -7.2     -0.03        -7.23        -0.02
```

TABLE 7 C

DIA.	RISE	RADIUS	ALPHA
m	m	m	deg.
14.300	1.783	15.228	28.00

h1	h2	S	b	d
mm	mm	m	mm	mm
75	180	2.787	250	700

y	db	qd	qb	p
mm	mm	kN/sq.m	kN/sq.m	m
-308	0	2.01	9.70	14.886

A	qD	ft	fb	Vol.
sq.m	kN/sq.m	N/sq.mm	N/sq.mm	cu.m
171	2.573	-0.194	0.422	18.6

s	Nt	Mf	Nf	Qf
m	kN/m	kNm/m	kN/m	kN/m
0.00	57.8	0.45	-4.25	-2.04
0.45	31.7	1.01	-6.88	-0.57
0.90	12.0	1.07	-8.47	0.23
1.35	-0.9	0.88	-9.23	0.55
1.80	-8.0	0.61	-9.40	0.59
2.25	-11.1	0.37	-9.22	0.49
2.70	-11.7	0.18	-8.87	0.34
3.15	-11.0	0.06	-8.49	0.21
3.60	-10.0	-0.01	-8.14	0.10
4.50	-8.2	-0.05	-7.66	-0.01
5.40	-7.4	-0.03	-7.46	-0.03

TABLE 8 C

DIA.	RISE	RADIUS	ALPHA
m	m	m	deg.
14.800	1.845	15.761	28.00

h1	h2	S	b	d
mm	mm	m	mm	mm
75	180	2.835	250	700

y	db	qd	qb	p
mm	mm	kN/sq.m	kN/sq.m	m
-308	0	2.01	9.06	15.406

A	qD	ft	fb	Vol.
sq.m	kN/sq.m	N/sq.mm	N/sq.mm	cu.m
183	2.561	-0.198	0.445	19.8

s	Nt	Mf	Nf	Qf
m	kN/m	kNm/m	kN/m	kN/m
0.00	61.1	0.50	-4.43	-2.10
0.45	33.8	1.08	-7.12	-0.59
0.90	13.1	1.14	-8.76	0.23
1.35	-0.5	0.94	-9.55	0.57
1.80	-8.1	0.67	-9.74	0.62
2.25	-11.5	0.41	-9.57	0.52
2.70	-12.2	0.21	-9.22	0.37
3.15	-11.6	0.07	-8.82	0.23
3.60	-10.5	-0.01	-8.46	0.12
4.50	-8.6	-0.05	-7.96	0.00
5.40	-7.7	-0.03	-7.74	-0.03

TABLE 9 C

DIA. m		RISE m		RADIUS m		ALPHA deg.
15.200		1.895		16.186		28.00

h1 mm	h2 mm	S m		b mm		d mm
75	180	2.873		250		700

y mm	db mm	qd kN/sq.m		qb kN/sq.m		p m
-308	0	2.00		8.59		15.822

A sq.m	qD kN/sq.m	ft N/sq.mm		fb N/sq.mm		Vol. cu.m
193	2.552	-0.201		0.464		20.8

s m	Nt kN/m	Mf kNm/m	Nf kN/m	Qf kN/m
0.00	63.8	0.55	-4.58	-2.14
0.45	35.6	1.13	-7.31	-0.61
0.90	14.1	1.20	-8.99	0.23
1.35	-0.1	1.00	-9.81	0.59
1.80	-8.1	0.71	-10.01	0.64
2.25	-11.8	0.44	-9.85	0.55
2.70	-12.7	0.23	-9.50	0.40
3.15	-12.1	0.08	-9.09	0.25
3.60	-11.0	0.00	-8.72	0.13
4.50	-8.9	-0.05	-8.19	0.00
5.85	-7.8	-0.02	-7.92	-0.03

TABLE 10 C

DIA.	RISE	RADIUS	ALPHA
m	m	m	deg.
15.700	1.957	16.719	28.00

h1	h2	S	b	d
mm	mm	m	mm	mm
75	180	2.920	250	700

y	db	qd	qb	p
mm	mm	kN/sq.m	kN/sq.m	m
-308	0	2.00	8.05	16.343

A	qD	ft	fb	Vol.
sq.m	kN/sq.m	N/sq.mm	N/sq.mm	cu.m
206	2.541	-0.203	0.488	22.1

s	Nt	Mf	Nf	Qf
m	kN/m	kNm/m	kN/m	kN/m
0.00	67.1	0.60	-4.76	-2.19
0.45	37.8	1.20	-7.55	-0.63
0.90	15.4	1.27	-9.28	0.24
1.35	0.5	1.06	-10.12	0.61
1.80	-8.2	0.77	-10.35	0.67
2.25	-12.1	0.48	-10.19	0.58
2.70	-13.2	0.26	-9.84	0.43
3.15	-12.7	0.10	-9.43	0.27
3.60	-11.5	0.00	-9.05	0.15
4.05	-10.3	-0.04	-8.73	0.06
4.95	-8.6	-0.05	-8.34	-0.02
5.85	-8.0	-0.03	-8.20	-0.03

TABLE 11 C

DIA.		RISE		RADIUS		ALPHA
m		m		m		deg.
16.100		2.007		17.145		28.00

h1	h2	S		b		d
mm	mm	m		mm		mm
75	180	2.957		250		700

y	db	qd		qb		p
mm	mm	kN/sq.m		kN/sq.m		m
-308	0	2.00		7.65		16.759

A	qD	ft		fb		Vol.
sq.m	kN/sq.m	N/sq.mm		N/sq.mm		cu.m
216	2.533	-0.205		0.507		23.2

s	Nt	Mf	Nf	Qf
m	kN/m	kNm/m	kN/m	kN/m
0.00	69.8	0.64	-4.90	-2.23
0.45	39.6	1.26	-7.74	-0.65
0.90	16.4	1.33	-9.50	0.24
1.35	0.9	1.12	-10.38	0.62
1.80	-8.1	0.81	-10.62	0.70
2.25	-12.4	0.52	-10.47	0.61
2.70	-13.6	0.28	-10.12	0.45
3.15	-13.2	0.11	-9.70	0.29
3.60	-12.0	0.01	-9.31	0.16
4.05	-10.7	-0.04	-8.98	0.07
4.95	-8.9	-0.06	-8.57	-0.02
5.85	-8.2	-0.03	-8.42	-0.03

TABLE 12 C

DIA. m	RISE m	RADIUS m	ALPHA deg.
16.600	2.070	17.677	28.00

h1 mm	h2 mm	S m	b mm	d mm
75	180	3.003	250	700

y mm	db mm	qd kN/sq.m	qb kN/sq.m	p m
-308	0	2.00	7.20	17.280

A sq.m	qD kN/sq.m	ft N/sq.mm	fb N/sq.mm	Vol. cu.m
230	2.523	-0.207	0.532	24.6

s m	Nt kN/m	Mf kNm/m	Nf kN/m	Qf kN/m
0.00	73.3	0.69	-5.09	-2.29
0.45	41.9	1.33	-7.98	-0.67
0.90	17.8	1.40	-9.78	0.24
1.35	1.5	1.19	-10.69	0.64
1.80	-8.1	0.87	-10.96	0.73
2.25	-12.8	0.56	-10.82	0.64
2.70	-14.2	0.31	-10.47	0.48
3.15	-13.8	0.13	-10.05	0.32
3.60	-12.6	0.02	-9.64	0.18
4.05	-11.3	-0.04	-9.30	0.08
4.95	-9.3	-0.06	-8.86	-0.02
6.30	-8.4	-0.02	-8.68	-0.03

TABLE 13 C

DIA.	RISE	RADIUS	ALPHA
m	m	m	deg.
17.100	2.132	18.210	28.00

h1	h2	S	b	d
mm	mm	m	mm	mm
75	180	3.047	250	700

y	db	qd	qb	p
mm	mm	kN/sq.m	kN/sq.m	m
-308	0	1.99	6.79	17.800

A	qD	ft	fb	Vol.
sq.m	kN/sq.m	N/sq.mm	N/sq.mm	cu.m
244	2.513	-0.209	0.556	26.0

s	Nt	Mf	Nf	Qf
m	kN/m	kNm/m	kN/m	kN/m
0.00	76.8	0.75	-5.27	-2.34
0.45	44.3	1.40	-8.22	-0.70
0.90	19.2	1.48	-10.07	0.24
1.35	2.1	1.26	-11.01	0.66
1.80	-8.0	0.94	-11.29	0.75
2.25	-13.1	0.61	-11.17	0.67
2.70	-14.7	0.34	-10.82	0.51
3.15	-14.4	0.15	-10.39	0.34
3.60	-13.2	0.03	-9.97	0.20
4.05	-11.8	-0.03	-9.62	0.10
5.40	-9.1	-0.05	-9.04	-0.03
6.30	-8.6	-0.03	-8.95	-0.03

TABLE 14 C

DIA. m	RISE m	RADIUS m	ALPHA deg.
17.600	2.194	18.742	28.00

h1 mm	h2 mm	S m	b mm	d mm
75	180	3.092	250	700

y mm	db mm	qd kN/sq.m	qb kN/sq.m	p m
-308	0	1.99	6.41	18.321

A sq.m	qD kN/sq.m	ft N/sq.mm	fb N/sq.mm	Vol. cu.m
258	2.504	-0.210	0.581	27.4

s m	Nt kN/m	Mf kNm/m	Nf kN/m	Qf kN/m
0.00	80.3	0.80	-5.45	-2.40
0.45	46.7	1.47	-8.46	-0.72
0.90	20.7	1.56	-10.34	0.24
1.35	2.8	1.34	-11.32	0.68
1.80	-7.9	1.00	-11.63	0.78
2.25	-13.4	0.66	-11.51	0.70
2.70	-15.2	0.38	-11.17	0.54
3.15	-15.0	0.17	-10.73	0.37
3.60	-13.8	0.04	-10.31	0.22
4.05	-12.4	-0.03	-9.94	0.11
4.50	-11.1	-0.06	-9.66	0.04
5.40	-9.4	-0.06	-9.33	-0.03
6.75	-8.9	-0.02	-9.22	-0.02

TABLE 15 C

DIA.	RISE	RADIUS	ALPHA
m	m	m	deg.
18.000	2.244	19.168	28.00

h1	h2	S	b	d
mm	mm	m	mm	mm
90	200	3.334	280	800

y	db	qd	qb	p
mm	mm	kN/sq.m	kN/sq.m	m
-349	0	2.37	8.82	18.737

A	qD	ft	fb	Vol.
sq.m	kN/sq.m	N/sq.mm	N/sq.mm	cu.m
270	2.929	-0.210	0.503	33.6

s	Nt	Mf	Nf	Qf
m	kN/m	kNm/m	kN/m	kN/m
0.00	77.3	0.76	-5.47	-2.51
0.45	47.2	1.50	-8.35	-0.89
0.90	23.1	1.66	-10.25	0.09
1.35	5.8	1.49	-11.34	0.59
1.80	-5.4	1.18	-11.79	0.76
2.25	-11.7	0.83	-11.81	0.74
2.70	-14.5	0.53	-11.57	0.61
3.15	-15.1	0.29	-11.19	0.45
3.60	-14.5	0.12	-10.78	0.30
4.05	-13.3	0.01	-10.39	0.18
4.50	-12.1	-0.05	-10.07	0.09
5.85	-9.6	-0.06	-9.51	-0.03
6.75	-9.2	-0.03	-9.41	-0.03

TABLE 16 C

DIA.	RISE		RADIUS	ALPHA
m	m		m	deg.
18.500	2.307		19.701	28.00
h1	h2	S	b	d
mm	mm	m	mm	mm
90	200	3.380	280	800
y	db	qd	qb	p
mm	mm	kN/sq.m	kN/sq.m	m
-349	0	2.36	8.35	19.258
A	qD	ft	fb	Vol.
sq.m	kN/sq.m	N/sq.mm	N/sq.mm	cu.m
286	2.920	-0.212	0.525	35.3
s	Nt	Mf	Nf	Qf
m	kN/m	kNm/m	kN/m	kN/m
0.00	80.8	0.82	-5.65	-2.56
0.45	49.6	1.58	-8.58	-0.92
0.90	24.7	1.74	-10.53	0.08
1.35	6.6	1.57	-11.65	0.60
1.80	-5.1	1.25	-12.12	0.79
2.25	-11.9	0.89	-12.16	0.77
2.70	-14.9	0.57	-11.92	0.64
3.15	-15.7	0.32	-11.54	0.48
3.60	-15.1	0.14	-11.12	0.33
4.05	-13.9	0.02	-10.72	0.20
4.50	-12.6	-0.04	-10.39	0.10
5.85	-9.9	-0.07	-9.80	-0.03
7.20	-9.4	-0.02	-9.67	-0.03

TABLE 17 C

DIA.	RISE	RADIUS	ALPHA
m	m	m	deg.
19.100	2.381	20.340	28.00

h1	h2	S	b	d
mm	mm	m	mm	mm
90	200	3.435	280	800

y	db	qd	qb	p
mm	mm	kN/sq.m	kN/sq.m	m
-349	0	2.36	7.83	19.882

A	qD	ft	fb	Vol.
sq.m	kN/sq.m	N/sq.mm	N/sq.mm	cu.m
304	2.909	-0.215	0.551	37.5

s	Nt	Mf	Nf	Qf
m	kN/m	kNm/m	kN/m	kN/m
0.00	84.9	0.89	-5.87	-2.62
0.45	52.6	1.67	-8.86	-0.95
0.90	26.6	1.84	-10.86	0.08
1.35	7.7	1.67	-12.01	0.62
1.80	-4.8	1.34	-12.52	0.82
2.25	-12.0	0.97	-12.57	0.80
2.70	-15.4	0.63	-12.33	0.68
3.15	-16.3	0.36	-11.95	0.52
3.60	-15.8	0.16	-11.53	0.36
4.05	-14.6	0.04	-11.12	0.22
4.50	-13.3	-0.04	-10.77	0.11
4.95	-12.0	-0.07	-10.49	0.04
5.85	-10.3	-0.07	-10.14	-0.03
7.20	-9.6	-0.03	-10.00	-0.03

```
TABLE  18  C
  DIA.         RISE         RADIUS        ALPHA
   m            m             m           deg.
 19.700        2.456        20.979        28.00
  h1        h2        S          b           d
  mm        mm        m          mm          mm
  90        200      3.488       280         800
   y        db        qd         qb           p
  mm        mm     kN/sq.m    kN/sq.m         m
 -349        0       2.36       7.36        20.507
   A        qD         ft         fb         Vol.
 sq.m    kN/sq.m   N/sq.mm    N/sq.mm        cu.m
  324     2.898     -0.216      0.578        39.8
   s        Nt         Mf         Nf           Qf
   m       kN/m      kNm/m       kN/m         kN/m
 0.00      89.2      0.96       -6.09        -2.69
 0.45      55.6      1.76       -9.13        -0.98
 0.90      28.5      1.95      -11.18         0.07
 1.35       8.7      1.77      -12.37         0.63
 1.80      -4.4      1.43      -12.91         0.84
 2.25     -12.1      1.04      -12.98         0.84
 2.70     -15.9      0.69      -12.75         0.72
 3.15     -17.0      0.40      -12.37         0.55
 3.60     -16.6      0.19      -11.94         0.39
 4.05     -15.4      0.05      -11.52         0.24
 4.50     -13.9     -0.03      -11.16         0.13
 4.95     -12.6     -0.07      -10.86         0.05
 6.30     -10.3     -0.06      -10.40        -0.04
 7.65      -9.9     -0.02      -10.33        -0.03
```

TABLE 19 C

DIA.	RISE	RADIUS	ALPHA
m	m	m	deg.
20.300	2.531	21.617	28.00

h1	h2	S	b	d
mm	mm	m	mm	mm
90	200	3.541	280	800

y	db	qd	qb	p
mm	mm	kN/sq.m	kN/sq.m	m
-349	0	2.35	6.93	21.131

A	qD	ft	fb	Vol.
sq.m	kN/sq.m	N/sq.mm	N/sq.mm	cu.m
344	2.888	-0.217	0.605	42.1

s	Nt	Mf	Nf	Qf
m	kN/m	kNm/m	kN/m	kN/m
0.00	93.4	1.04	-6.31	-2.75
0.45	58.7	1.86	-9.41	-1.02
0.90	30.6	2.05	-11.50	0.07
1.35	9.9	1.87	-12.74	0.65
1.80	-4.0	1.52	-13.30	0.87
2.25	-12.2	1.12	-13.38	0.88
2.70	-16.3	0.75	-13.17	0.76
3.15	-17.7	0.45	-12.79	0.59
3.60	-17.3	0.22	-12.35	0.42
4.05	-16.1	0.07	-11.92	0.27
4.50	-14.6	-0.03	-11.54	0.15
4.95	-13.2	-0.07	-11.24	0.06
6.30	-10.6	-0.07	-10.74	-0.04
7.65	-10.2	-0.02	-10.65	-0.03

TABLE 20 C

DIA.	RISE	RADIUS	ALPHA
m	m	m	deg.
20.900	2.606	22.256	28.00

h1	h2	S	b	d
mm	mm	m	mm	mm
90	200	3.593	280	800

y	db	qd	qb	p
mm	mm	kN/sq.m	kN/sq.m	m
-349	0	2.35	6.54	21.756

A	qD	ft	fb	Vol.
sq.m	kN/sq.m	N/sq.mm	N/sq.mm	cu.m
364	2.878	-0.218	0.632	44.4

s	Nt	Mf	Nf	Qf
m	kN/m	kNm/m	kN/m	kN/m
0.00	97.8	1.11	-6.52	-2.82
0.45	61.8	1.95	-9.68	-1.05
0.90	32.7	2.16	-11.82	0.06
1.35	11.1	1.98	-13.09	0.66
1.80	-3.5	1.62	-13.69	0.90
2.25	-12.3	1.20	-13.79	0.91
2.70	-16.8	0.82	-13.58	0.80
3.15	-18.3	0.49	-13.21	0.63
3.60	-18.1	0.25	-12.76	0.45
4.05	-16.9	0.09	-12.32	0.29
4.50	-15.3	-0.02	-11.93	0.17
4.95	-13.9	-0.07	-11.61	0.08
5.40	-12.6	-0.09	-11.37	0.02
6.75	-10.7	-0.06	-11.01	-0.04
8.10	-10.5	-0.02	-10.99	-0.02

TABLE 21 C

DIA.		RISE		RADIUS		ALPHA
m		m		m		deg.
21.500		2.681		22.895		28.00

h1	h2		S		b	d
mm	mm		m		mm	mm
90	200		3.644		280	800

y	db	qd		qb		p
mm	mm	kN/sq.m		kN/sq.m		m
-349	0	2.35		6.18		22.380

A	qD	ft	fb	Vol.
sq.m	kN/sq.m	N/sq.mm	N/sq.mm	cu.m
386	2.869	-0.218	0.660	46.9

s	Nt	Mf	Nf	Qf
m	kN/m	kNm/m	kN/m	kN/m
0.00	102.2	1.19	-6.74	-2.88
0.45	65.0	2.05	-9.96	-1.09
0.90	34.8	2.26	-12.14	0.05
1.35	12.3	2.08	-13.45	0.67
1.80	-3.0	1.71	-14.07	0.93
2.25	-12.3	1.29	-14.20	0.95
2.70	-17.2	0.88	-14.00	0.84
3.15	-19.0	0.54	-13.62	0.66
3.60	-18.8	0.29	-13.17	0.48
4.05	-17.6	0.11	-12.73	0.32
4.50	-16.1	-0.01	-12.32	0.19
4.95	-14.5	-0.07	-11.99	0.09
5.40	-13.2	-0.09	-11.73	0.03
6.75	-11.0	-0.07	-11.35	-0.04
8.10	-10.8	-0.02	-11.31	-0.03

TABLE 22 C

DIA.	RISE	RADIUS	ALPHA
m	m	m	deg.
22.100	2.755	23.534	28.00

h1	h2	S	b	d
mm	mm	m	mm	mm
90	200	3.695	280	800

y	db	qd	qb	p
mm	mm	kN/sq.m	kN/sq.m	m
-349	0	2.34	5.85	23.005

A	qD	ft	fb	Vol.
sq.m	kN/sq.m	N/sq.mm	N/sq.mm	cu.m
408	2.860	-0.218	0.688	49.4

s	Nt	Mf	Nf	Qf
m	kN/m	kNm/m	kN/m	kN/m
0.00	106.6	1.26	-6.96	-2.95
0.45	68.3	2.15	-10.23	-1.12
0.90	37.0	2.37	-12.46	0.04
1.35	13.6	2.19	-13.81	0.68
1.80	-2.4	1.81	-14.46	0.95
2.25	-12.3	1.37	-14.60	0.98
2.70	-17.6	0.95	-14.41	0.87
3.15	-19.6	0.60	-14.04	0.70
3.60	-19.6	0.32	-13.59	0.52
4.05	-18.4	0.13	-13.13	0.35
4.50	-16.8	0.00	-12.72	0.21
4.95	-15.2	-0.07	-12.37	0.11
5.40	-13.8	-0.10	-12.10	0.03
6.75	-11.4	-0.07	-11.68	-0.04
8.55	-11.1	-0.01	-11.65	-0.02

TABLE 23 C

DIA.		RISE		RADIUS		ALPHA
m		m		m		deg.
22.800		2.843		24.280		28.00

h1	h2	S	b	d
mm	mm	m	mm	mm
90	200	3.753	280	800

y	db	qd	qb	p
mm	mm	kN/sq.m	kN/sq.m	m
-349	0	2.34	5.50	23.734

A	qD	ft	fb	Vol.
sq.m	kN/sq.m	N/sq.mm	N/sq.mm	cu.m
434	2.850	-0.217	0.720	52.4

s	Nt	Mf	Nf	Qf
m	kN/m	kNm/m	kN/m	kN/m
0.00	111.8	1.35	-7.21	-3.02
0.45	72.1	2.26	-10.54	-1.17
0.90	39.7	2.50	-12.83	0.02
1.35	15.2	2.32	-14.22	0.69
1.80	-1.7	1.93	-14.90	0.98
2.25	-12.2	1.47	-15.07	1.02
2.70	-18.0	1.03	-14.90	0.92
3.15	-20.3	0.66	-14.53	0.75
3.60	-20.4	0.37	-14.07	0.56
4.05	-19.3	0.16	-13.60	0.38
4.50	-17.7	0.02	-13.18	0.24
4.95	-16.0	-0.06	-12.81	0.12
5.40	-14.5	-0.10	-12.53	0.05
5.85	-13.2	-0.11	-12.31	0.00
7.20	-11.5	-0.06	-12.03	-0.04
8.55	-11.4	-0.02	-12.03	-0.02

```
TABLE  24  C
  DIA.          RISE          RADIUS        ALPHA
   m             m              m            deg.
 23.500        2.930         25.025         28.00
   h1        h2        S           b           d
   mm        mm        m           mm          mm
   90        200      3.810        280         800
    y        db        qd          qb           p
   mm        mm    kN/sq.m     kN/sq.m          m
 -349         0      2.34         5.17       24.462
    A         qD        ft          fb         Vol.
  sq.m    kN/sq.m   N/sq.mm     N/sq.mm        cu.m
  461      2.840    -0.215       0.754         55.4
    s        Nt        Mf          Nf           Qf
    m       kN/m     kNm/m        kN/m         kN/m
  0.00     117.2     1.43        -7.46        -3.10
  0.45      76.1     2.38       -10.85        -1.21
  0.90      42.4     2.62       -13.19         0.01
  1.35      16.9     2.45       -14.63         0.70
  1.80      -0.8     2.05       -15.35         1.01
  2.25     -12.1     1.58       -15.54         1.06
  2.70     -18.3     1.12       -15.38         0.96
  3.15     -21.0     0.73       -15.02         0.79
  3.60     -21.3     0.41       -14.56         0.60
  4.05     -20.3     0.19       -14.08         0.42
  4.50     -18.6     0.03       -13.64         0.26
  4.95     -16.8    -0.06       -13.26         0.14
  5.40     -15.2    -0.10       -12.96         0.06
  5.85     -13.9    -0.11       -12.73         0.00
  7.20     -11.8    -0.07       -12.41        -0.05
  9.00     -11.8    -0.01       -12.42        -0.02
```

TABLE 25 C

DIA.		RISE		RADIUS		ALPHA	
m		m		m		deg.	
24.000		2.992		25.558		28.00	

h1	h2	S	b	d
mm	mm	m	mm	mm
100	250	4.230	300	1000

y	db	qd	qb	p
mm	mm	kN/sq.m	kN/sq.m	m
-443	-0	2.68	6.12	24.983

A	qD	ft	fb	Vol.
sq.m	kN/sq.m	N/sq.mm	N/sq.mm	cu.m
481	3.411	-0.222	0.584	69.5

s	Nt	Mf	Nf	Qf
m	kN/m	kNm/m	kN/m	kN/m
0.00	112.1	1.28	-6.98	-3.51
0.45	77.8	2.42	-10.22	-1.68
0.90	48.5	2.87	-12.62	-0.41
1.35	24.9	2.86	-14.26	0.40
1.80	7.2	2.57	-15.25	0.84
2.25	-5.3	2.14	-15.74	1.03
2.70	-13.4	1.67	-15.84	1.04
3.15	-18.1	1.22	-15.68	0.94
3.60	-20.3	0.83	-15.36	0.79
4.05	-20.7	0.51	-14.95	0.62
4.50	-20.1	0.27	-14.52	0.46
4.95	-18.9	0.09	-14.09	0.31
5.40	-17.5	-0.02	-13.72	0.20
5.85	-16.1	-0.09	-13.39	0.10
7.65	-12.6	-0.10	-12.67	-0.04
9.00	-12.1	-0.04	-12.57	-0.04

```
TABLE  26  C
```

DIA. m		RISE m		RADIUS m		ALPHA deg.
24.700		3.080		26.303		28.00

h1 mm	h2 mm	S m	b mm	d mm
100	250	4.291	300	1000

y mm	db mm	qd kN/sq.m	qb kN/sq.m	p m
-443	-0	2.67	5.78	25.712

A sq.m	qD kN/sq.m	ft N/sq.mm	fb N/sq.mm	Vol. cu.m
509	3.398	-0.224	0.610	73.3

s m	Nt kN/m	Mf kNm/m	Nf kN/m	Qf kN/m
0.00	117.2	1.38	-7.22	-3.59
0.45	81.8	2.55	-10.52	-1.73
0.90	51.4	3.02	-12.96	-0.44
1.35	26.9	3.01	-14.65	0.40
1.80	8.4	2.72	-15.68	0.86
2.25	-4.8	2.28	-16.19	1.06
2.70	-13.4	1.79	-16.32	1.08
3.15	-18.5	1.32	-16.17	0.99
3.60	-20.9	0.91	-15.85	0.84
4.05	-21.5	0.57	-15.44	0.66
4.50	-21.0	0.31	-15.00	0.50
4.95	-19.8	0.12	-14.56	0.34
5.40	-18.3	-0.01	-14.17	0.22
5.85	-16.9	-0.08	-13.84	0.12
6.30	-15.6	-0.12	-13.56	0.05
7.65	-13.1	-0.11	-13.07	-0.04
9.45	-12.4	-0.03	-12.95	-0.03

```
TABLE  27  C
```

DIA.	RISE	RADIUS	ALPHA
m	m	m	deg.
25.500	3.179	27.155	28.00

h1	h2	S	b	d
mm	mm	m	mm	mm
100	250	4.360	300	1000

y	db	qd	qb	p
mm	mm	kN/sq.m	kN/sq.m	m
-443	-0	2.67	5.42	26.544

A	qD	ft	fb	Vol.
sq.m	kN/sq.m	N/sq.mm	N/sq.mm	cu.m
543	3.383	-0.224	0.640	77.8

s	Nt	Mf	Nf	Qf
m	kN/m	kNm/m	kN/m	kN/m
0.00	123.2	1.49	-7.50	-3.68
0.45	86.5	2.70	-10.86	-1.79
0.90	54.8	3.19	-13.36	-0.46
1.35	29.3	3.19	-15.09	0.39
1.80	9.8	2.89	-16.16	0.88
2.25	-4.1	2.43	-16.71	1.10
2.70	-13.4	1.93	-16.85	1.13
3.15	-18.9	1.43	-16.72	1.04
3.60	-21.6	1.00	-16.41	0.89
4.05	-22.4	0.64	-15.99	0.71
4.50	-22.0	0.36	-15.54	0.54
4.95	-20.8	0.15	-15.10	0.38
5.40	-19.3	0.01	-14.70	0.25
5.85	-17.8	-0.08	-14.35	0.14
6.30	-16.4	-0.12	-14.06	0.07
8.10	-13.2	-0.10	-13.45	-0.05
9.90	-12.9	-0.03	-13.39	-0.03

TABLE 28 C

DIA.	RISE	RADIUS	ALPHA
m	m	m	deg.
26.200	3.267	27.900	28.00
h1	h2	S	b	d
mm	mm	m	mm	mm
100	250	4.419	300	1000
y	db	qd	qb	p
mm	mm	kN/sq.m	kN/sq.m	m
-443	-0	2.66	5.14	27.273
A	qD	ft	fb	Vol.
sq.m	kN/sq.m	N/sq.mm	N/sq.mm	cu.m
573	3.371	-0.224	0.667	81.8

s	Nt	Mf	Nf	Qf
m	kN/m	kNm/m	kN/m	kN/m
0.00	128.5	1.59	-7.74	-3.77
0.45	90.6	2.83	-11.15	-1.84
0.90	57.9	3.33	-13.70	-0.49
1.35	31.4	3.34	-15.48	0.39
1.80	11.1	3.04	-16.58	0.90
2.25	-3.5	2.57	-17.16	1.13
2.70	-13.3	2.05	-17.32	1.17
3.15	-19.2	1.54	-17.20	1.08
3.60	-22.2	1.08	-16.89	0.93
4.05	-23.2	0.70	-16.48	0.76
4.50	-22.8	0.40	-16.03	0.58
4.95	-21.7	0.18	-15.58	0.41
5.40	-20.2	0.03	-15.16	0.27
5.85	-18.6	-0.07	-14.80	0.16
6.30	-17.1	-0.12	-14.49	0.08
6.75	-15.9	-0.14	-14.25	0.02
8.10	-13.6	-0.11	-13.85	-0.05
9.90	-13.2	-0.03	-13.77	-0.03

TABLE 29 C

DIA.		RISE		RADIUS		ALPHA
m		m		m		deg.
27.000		3.366		28.752		28.00
h1	h2		S	b		d
mm	mm		m	mm		mm
100	250		4.486	300		1000
y	db	qd		qb		p
mm	mm	kN/sq.m		kN/sq.m		m
-443	-0	2.66		4.84		28.106
A	qD	ft		fb		Vol.
sq.m	kN/sq.m	N/sq.mm		N/sq.mm		cu.m
608	3.358	-0.224		0.697		86.5

s	Nt	Mf	Nf	Qf
m	kN/m	kNm/m	kN/m	kN/m
0.00	134.6	1.70	-8.01	-3.86
0.45	95.4	2.98	-11.49	-1.91
0.90	61.5	3.50	-14.09	-0.52
1.35	33.9	3.52	-15.91	0.38
1.80	12.7	3.21	-17.06	0.91
2.25	-2.7	2.74	-17.67	1.16
2.70	-13.1	2.20	-17.86	1.21
3.15	-19.5	1.66	-17.75	1.13
3.60	-22.9	1.19	-17.45	0.99
4.05	-24.1	0.78	-17.04	0.80
4.50	-23.8	0.46	-16.58	0.62
4.95	-22.7	0.22	-16.12	0.45
5.40	-21.2	0.05	-15.69	0.31
5.85	-19.5	-0.06	-15.31	0.19
6.30	-18.0	-0.12	-14.99	0.10
6.75	-16.6	-0.15	-14.74	0.03
8.55	-13.7	-0.10	-14.24	-0.05
10.35	-13.6	-0.02	-14.22	-0.03

TABLE 30 C

DIA. m		RISE m		RADIUS m		ALPHA deg.
27.800		3.466		29.604		28.00

h1 mm	h2 mm	S m	b mm	d mm
100	250	4.552	300	1000

y mm	db mm	qd kN/sq.m	qb kN/sq.m	p m
-443	-0	2.66	4.56	28.938

A sq.m	qD kN/sq.m	ft N/sq.mm	fb N/sq.mm	Vol. cu.m
645	3.345	-0.223	0.729	91.4

s m	Nt kN/m	Mf kNm/m	Nf kN/m	Qf kN/m
0.00	140.8	1.81	-8.28	-3.96
0.45	100.3	3.12	-11.82	-1.97
0.90	65.2	3.67	-14.48	-0.56
1.35	36.6	3.70	-16.35	0.38
1.80	14.4	3.39	-17.54	0.93
2.25	-1.8	2.90	-18.18	1.19
2.70	-12.9	2.35	-18.39	1.25
3.15	-19.8	1.79	-18.30	1.18
3.60	-23.5	1.29	-18.01	1.04
4.05	-25.0	0.87	-17.60	0.85
4.50	-24.8	0.52	-17.13	0.67
4.95	-23.8	0.26	-16.66	0.49
5.40	-22.2	0.08	-16.22	0.34
5.85	-20.5	-0.04	-15.83	0.21
6.30	-18.9	-0.12	-15.50	0.12
6.75	-17.4	-0.15	-15.23	0.05
7.20	-16.2	-0.16	-15.02	0.00
9.00	-13.9	-0.09	-14.65	-0.05
10.80	-14.0	-0.02	-14.67	-0.02

TABLE 31 C

DIA.		RISE		RADIUS		ALPHA
m		m		m		deg.
28.700		3.578		30.563		28.00
h1	h2	S		b		d
mm	mm	m		mm		mm
100	250	4.625		300		1000
y	db	qd		qb		p
mm	mm	kN/sq.m		kN/sq.m		m
-443	-0	2.65		4.28		29.875
A	qD	ft		fb		Vol.
sq.m	kN/sq.m	N/sq.mm		N/sq.mm		cu.m
687	3.331	-0.221		0.764		97.0

s	Nt	Mf	Nf	Qf
m	kN/m	kNm/m	kN/m	kN/m
0.00	147.9	1.94	-8.59	-4.06
0.45	105.9	3.29	-12.19	-2.04
0.90	69.5	3.87	-14.91	-0.59
1.35	39.6	3.90	-16.84	0.37
1.80	16.4	3.59	-18.07	0.94
2.25	-0.7	3.09	-18.74	1.23
2.70	-12.6	2.52	-18.99	1.30
3.15	-20.1	1.94	-18.92	1.24
3.60	-24.2	1.42	-18.63	1.09
4.05	-25.9	0.96	-18.22	0.91
4.50	-26.0	0.60	-17.76	0.72
4.95	-25.0	0.32	-17.28	0.54
5.40	-23.4	0.11	-16.82	0.38
5.85	-21.6	-0.03	-16.42	0.24
6.30	-19.9	-0.11	-16.07	0.14
6.75	-18.3	-0.16	-15.78	0.06
7.20	-17.0	-0.17	-15.56	0.01
9.00	-14.4	-0.10	-15.14	-0.06
10.80	-14.4	-0.02	-15.15	-0.03

TABLE 32 C

DIA.		RISE		RADIUS		ALPHA
m		m		m		deg.
29.500		3.678		31.415		28.00
h1	h2		S	b		d
mm	mm		m	mm		mm
100	250		4.689	300		1000
y	db	qd		qb		p
mm	mm	kN/sq.m		kN/sq.m		m
-443	-0	2.65		4.05		30.708
A	qD		ft	fb		Vol.
sq.m	kN/sq.m		N/sq.mm	N/sq.mm		cu.m
726	3.319		-0.219	0.796		102.1
s	Nt		Mf	Nf		Qf
m	kN/m		kNm/m	kN/m		kN/m
0.00	154.2		2.06	-8.86		-4.16
0.45	111.0		3.44	-12.52		-2.11
0.90	73.4		4.04	-15.30		-0.63
1.35	42.4		4.08	-17.26		0.36
1.80	18.2		3.77	-18.54		0.96
2.25	0.3		3.27	-19.25		1.26
2.70	-12.2		2.68	-19.51		1.34
3.15	-20.2		2.08	-19.46		1.29
3.60	-24.8		1.53	-19.19		1.15
4.05	-26.8		1.06	-18.78		0.96
4.50	-27.0		0.67	-18.31		0.76
4.95	-26.0		0.37	-17.83		0.58
5.40	-24.5		0.14	-17.36		0.41
5.85	-22.7		-0.01	-16.94		0.27
6.30	-20.8		-0.10	-16.58		0.16
6.75	-19.2		-0.16	-16.28		0.08
7.20	-17.8		-0.18	-16.04		0.02
7.65	-16.7		-0.17	-15.86		-0.02
9.45	-14.6		-0.09	-15.56		-0.05
11.25	-14.9		-0.01	-15.60		-0.02

TABLE 33 C

DIA. m		RISE m	RADIUS m	ALPHA deg.
30.000		3.740	31.947	28.00

h1 mm	h2 mm	S m	b mm	d mm
130	290	5.180	300	1200

y mm	db mm	qd kN/sq.m	qb kN/sq.m	p m
-526	-0	3.40	6.62	31.229

A sq.m	qD kN/sq.m	ft N/sq.mm	fb N/sq.mm	Vol. cu.m
751	4.169	-0.224	0.661	132.7

s m	Nt kN/m	Mf kNm/m	Nf kN/m	Qf kN/m
0.00	148.9	1.83	-8.37	-4.57
0.45	111.5	3.42	-11.88	-2.58
0.90	78.2	4.23	-14.66	-1.09
1.35	49.8	4.46	-16.74	-0.03
1.80	26.6	4.31	-18.21	0.68
2.25	8.5	3.90	-19.15	1.09
2.70	-5.0	3.36	-19.66	1.29
3.15	-14.5	2.76	-19.83	1.33
3.60	-20.7	2.18	-19.75	1.26
4.05	-24.4	1.64	-19.49	1.13
4.50	-26.0	1.17	-19.12	0.96
4.95	-26.3	0.77	-18.69	0.78
5.40	-25.7	0.46	-18.23	0.61
5.85	-24.5	0.22	-17.79	0.45
6.30	-23.1	0.05	-17.38	0.32
6.75	-21.5	-0.07	-17.01	0.21
7.20	-20.1	-0.14	-16.70	0.12
7.65	-18.8	-0.18	-16.43	0.05
9.45	-15.6	-0.15	-15.84	-0.06
11.70	-15.1	-0.03	-15.75	-0.03

TABLE 34 C

DIA. m	RISE m	RADIUS m	ALPHA deg.
30.900	3.853	32.905	28.00

h1 mm	h2 mm	S m	b mm	d mm
130	290	5.257	300	1200

y mm	db mm	qd kN/sq.m	qb kN/sq.m	p m
-526	-0	3.39	6.24	32.165

A sq.m	qD kN/sq.m	ft N/sq.mm	fb N/sq.mm	Vol. cu.m
797	4.155	-0.224	0.691	140.3

s m	Nt kN/m	Mf kNm/m	Nf kN/m	Qf kN/m
0.00	155.9	1.96	-8.66	-4.69
0.45	117.2	3.60	-12.23	-2.67
0.90	82.7	4.44	-15.07	-1.14
1.35	53.2	4.69	-17.20	-0.05
1.80	29.1	4.54	-18.71	0.68
2.25	10.1	4.12	-19.69	1.11
2.70	-4.1	3.57	-20.23	1.33
3.15	-14.2	2.95	-20.43	1.38
3.60	-20.9	2.34	-20.36	1.32
4.05	-25.0	1.78	-20.11	1.19
4.50	-26.9	1.28	-19.74	1.02
4.95	-27.4	0.87	-19.31	0.84
5.40	-26.8	0.53	-18.85	0.66
5.85	-25.7	0.27	-18.40	0.50
6.30	-24.2	0.08	-17.98	0.35
6.75	-22.6	-0.05	-17.60	0.24
7.20	-21.1	-0.14	-17.27	0.14
7.65	-19.7	-0.18	-16.99	0.07
9.90	-15.8	-0.14	-16.29	-0.06
11.70	-15.5	-0.04	-16.24	-0.04

TABLE 35 C

DIA.	RISE		RADIUS	ALPHA
m	m		m	deg.
31.800	3.965		33.864	28.00
h1	h2	S	b	d
mm	mm	m	mm	mm
130	290	5.333	300	1200
y	db	qd	qb	p
mm	mm	kN/sq.m	kN/sq.m	m
-526	-0	3.39	5.89	33.102
A	qD	ft	fb	Vol.
sq.m	kN/sq.m	N/sq.mm	N/sq.mm	cu.m
844	4.141	-0.223	0.721	148.0
s	Nt	Mf	Nf	Qf
m	kN/m	kNm/m	kN/m	kN/m
0.00	162.9	2.10	-8.95	-4.81
0.45	123.1	3.78	-12.59	-2.75
0.90	87.4	4.65	-15.47	-1.19
1.35	56.8	4.92	-17.66	-0.07
1.80	31.6	4.77	-19.22	0.68
2.25	11.8	4.35	-20.23	1.13
2.70	-3.2	3.78	-20.80	1.36
3.15	-13.9	3.15	-21.02	1.42
3.60	-21.1	2.52	-20.97	1.37
4.05	-25.5	1.93	-20.74	1.24
4.50	-27.8	1.41	-20.37	1.08
4.95	-28.4	0.96	-19.94	0.89
5.40	-28.0	0.60	-19.47	0.71
5.85	-26.8	0.32	-19.01	0.54
6.30	-25.3	0.11	-18.58	0.39
6.75	-23.7	-0.03	-18.18	0.27
7.20	-22.1	-0.13	-17.84	0.16
7.65	-20.6	-0.18	-17.55	0.09
8.10	-19.3	-0.21	-17.31	0.03
9.90	-16.3	-0.15	-16.79	-0.06
12.15	-16.0	-0.03	-16.74	-0.03

TABLE 36 C

DIA.	RISE	RADIUS	ALPHA
m	m	m	deg.
32.800	4.090	34.929	28.00

h1	h2	S	b	d
mm	mm	m	mm	mm
130	290	5.417	300	1200

y	db	qd	qb	p
mm	mm	kN/sq.m	kN/sq.m	m
-526	-0	3.39	5.54	34.143

A	qD	ft	fb	Vol.
sq.m	kN/sq.m	N/sq.mm	N/sq.mm	cu.m
898	4.126	-0.221	0.756	157.0

s	Nt	Mf	Nf	Qf
m	kN/m	kNm/m	kN/m	kN/m
0.00	170.9	2.25	-9.27	-4.94
0.45	129.7	3.98	-12.97	-2.84
0.90	92.6	4.88	-15.92	-1.25
1.35	60.8	5.17	-18.16	-0.10
1.80	34.6	5.03	-19.77	0.67
2.25	13.8	4.61	-20.83	1.15
2.70	-2.0	4.03	-21.43	1.40
3.15	-13.5	3.38	-21.68	1.48
3.60	-21.3	2.72	-21.65	1.43
4.05	-26.1	2.10	-21.43	1.31
4.50	-28.7	1.55	-21.07	1.14
4.95	-29.5	1.08	-20.63	0.95
5.40	-29.2	0.69	-20.17	0.77
5.85	-28.2	0.39	-19.70	0.59
6.30	-26.6	0.16	-19.25	0.43
6.75	-25.0	-0.01	-18.84	0.30
7.20	-23.3	-0.12	-18.48	0.19
7.65	-21.7	-0.18	-18.17	0.11
8.10	-20.3	-0.22	-17.92	0.04
10.35	-16.6	-0.14	-17.30	-0.06
12.60	-16.5	-0.03	-17.29	-0.03

TABLE 37 C

DIA.	RISE	RADIUS	ALPHA
m	m	m	deg.
33.800	4.214	35.994	28.00

h1	h2	S	b	d
mm	mm	m	mm	mm
130	290	5.499	300	1200

y	db	qd	qb	p
mm	mm	kN/sq.m	kN/sq.m	m
-526	-0	3.38	5.22	35.184

A	qD	ft	fb	Vol.
sq.m	kN/sq.m	N/sq.mm	N/sq.mm	cu.m
953	4.112	-0.219	0.790	166.1

s	Nt	Mf	Nf	Qf
m	kN/m	kNm/m	kN/m	kN/m
0.00	179.0	2.40	-9.59	-5.07
0.45	136.4	4.18	-13.36	-2.94
0.90	98.0	5.12	-16.37	-1.32
1.35	65.0	5.43	-18.66	-0.14
1.80	37.7	5.30	-20.32	0.67
2.25	15.9	4.87	-21.42	1.17
2.70	-0.8	4.28	-22.06	1.44
3.15	-13.0	3.61	-22.34	1.53
3.60	-21.3	2.92	-22.33	1.49
4.05	-26.7	2.28	-22.12	1.37
4.50	-29.6	1.70	-21.77	1.20
4.95	-30.7	1.20	-21.33	1.01
5.40	-30.5	0.79	-20.86	0.82
5.85	-29.5	0.46	-20.38	0.64
6.30	-28.0	0.21	-19.92	0.48
6.75	-26.2	0.02	-19.50	0.34
7.20	-24.5	-0.10	-19.12	0.22
7.65	-22.8	-0.18	-18.80	0.13
8.10	-21.3	-0.22	-18.53	0.06
8.55	-20.0	-0.23	-18.31	0.01
10.80	-16.9	-0.13	-17.82	-0.06
13.05	-17.0	-0.02	-17.85	-0.03

```
TABLE  38  C
   DIA.          RISE          RADIUS         ALPHA
    m             m              m             deg.
 34.800         4.339         37.059         28.00
   h1       h2        S          b             d
   mm       mm        m          mm            mm
  130       290      5.579      300           1200
   y        db        qd         qb            p
   mm       mm     kN/sq.m    kN/sq.m          m
 -526       -0       3.38       4.92          36.225
    A        qD        ft         fb          Vol.
  sq.m    kN/sq.m   N/sq.mm    N/sq.mm        cu.m
  1010     4.099    -0.216     0.826         175.5
    s        Nt        Mf         Nf            Qf
    m       kN/m     kNm/m      kN/m          kN/m
  0.00     187.2     2.55       -9.91         -5.20
  0.45     143.2     4.38      -13.74         -3.04
  0.90     103.6     5.36      -16.81         -1.38
  1.35      69.4     5.69      -19.16         -0.17
  1.80      40.9     5.57      -20.86          0.66
  2.25      18.1     5.14      -22.01          1.19
  2.70       0.6     4.53      -22.69          1.47
  3.15     -12.3     3.84      -22.99          1.57
  3.60     -21.3     3.14      -23.00          1.55
  4.05     -27.1     2.46      -22.80          1.43
  4.50     -30.4     1.86      -22.46          1.27
  4.95     -31.7     1.33      -22.03          1.08
  5.40     -31.7     0.89      -21.55          0.88
  5.85     -30.8     0.53      -21.07          0.69
  6.30     -29.3     0.26      -20.60          0.52
  6.75     -27.5     0.06      -20.16          0.38
  7.20     -25.7    -0.08      -19.77          0.25
  7.65     -23.9    -0.17      -19.43          0.15
  8.10     -22.3    -0.22      -19.15          0.08
  8.55     -20.9    -0.24      -18.91          0.02
  9.00     -19.8    -0.24      -18.73         -0.02
 11.25     -17.3    -0.11      -18.35         -0.06
 13.50     -17.6    -0.01      -18.41         -0.02
```

TABLE 39 C

DIA.	RISE	RADIUS	ALPHA
m	m	m	deg.
35.000	4.364	37.272	28.00

h1	h2	S	b	d
mm	mm	m	mm	mm
130	290	5.595	300	1200

y	db	qd	qb	p
mm	mm	kN/sq.m	kN/sq.m	m
-526	-0	3.38	4.87	36.433

A	qD	ft	fb	Vol.
sq.m	kN/sq.m	N/sq.mm	N/sq.mm	cu.m
1022	4.096	-0.215	0.833	177.4

s	Nt	Mf	Nf	Qf
m	kN/m	kNm/m	kN/m	kN/m
0.00	188.9	2.58	-9.97	-5.22
0.45	144.6	4.42	-13.82	-3.06
0.90	104.7	5.41	-16.90	-1.39
1.35	70.2	5.74	-19.26	-0.18
1.80	41.6	5.62	-20.97	0.66
2.25	18.6	5.19	-22.12	1.19
2.70	0.9	4.59	-22.81	1.48
3.15	-12.2	3.89	-23.12	1.58
3.60	-21.3	3.18	-23.14	1.56
4.05	-27.2	2.50	-22.94	1.45
4.50	-30.6	1.89	-22.60	1.28
4.95	-32.0	1.35	-22.17	1.09
5.40	-32.0	0.91	-21.69	0.89
5.85	-31.1	0.55	-21.20	0.70
6.30	-29.6	0.27	-20.73	0.53
6.75	-27.8	0.07	-20.29	0.38
7.20	-26.0	-0.08	-19.90	0.26
7.65	-24.2	-0.17	-19.56	0.16
8.10	-22.5	-0.22	-19.27	0.08
8.55	-21.1	-0.25	-19.04	0.02
9.00	-20.0	-0.25	-18.85	-0.02
11.25	-17.4	-0.12	-18.46	-0.06
13.50	-17.7	-0.02	-18.51	-0.03

```
TABLE  40  C
  DIA.           RISE          RADIUS         ALPHA
   m              m              m             deg.
 36.100         4.501         38.443         28.00
  h1            h2        S          b            d
  mm            mm        m          mm           mm
 130           290       5.683      300          1200
  y            db        qd         qb            p
  mm           mm     kN/sq.m    kN/sq.m          m
-526          -0        3.37       4.57         37.578
  A            qD        ft         fb           Vol.
 sq.m      kN/sq.m    N/sq.mm    N/sq.mm        cu.m
1087         4.082     -0.211      0.872        188.1
  s            Nt        Mf         Nf           Qf
  m           kN/m      kNm/m      kN/m         kN/m
 0.00        198.0      2.75      -10.32        -5.37
 0.45        152.2      4.65      -14.24        -3.16
 0.90        110.9      5.67      -17.38        -1.47
 1.35         75.1      6.03      -19.80        -0.22
 1.80         45.3      5.92      -21.57         0.65
 2.25         21.2      5.49      -22.77         1.21
 2.70          2.5      4.87      -23.49         1.51
 3.15        -11.4      4.16      -23.83         1.63
 3.60        -21.2      3.42      -23.88         1.62
 4.05        -27.7      2.71      -23.70         1.51
 4.50        -31.4      2.07      -23.36         1.35
 4.95        -33.1      1.50      -22.94         1.16
 5.40        -33.3      1.03      -22.46         0.96
 5.85        -32.5      0.64      -21.96         0.76
 6.30        -31.1      0.34      -21.48         0.59
 6.75        -29.3      0.11      -21.02         0.43
 7.20        -27.4     -0.05      -20.62         0.30
 7.65        -25.5     -0.16      -20.26         0.19
 8.10        -23.7     -0.22      -19.96         0.10
 8.55        -22.2     -0.25      -19.70         0.04
 9.00        -20.9     -0.26      -19.50        -0.01
11.70        -17.8     -0.10      -19.05        -0.06
13.95        -18.3     -0.01      -19.12        -0.02
```

Appendix 6: Design Tables for Prestressed Domes

Tables 1AP to 40AP   For Type A Domes

TABLE 1 AP

DIA. m		RISE m	RADIUS m	ALPHA deg.
12.000		1.496	12.779	28.00

h1 mm	h2 mm	S m	b mm	d mm
75	180	2.553	250	700

y mm	db mm	qd kN/sq.m	qb kN/sq.m	p m
189	496	2.03	13.78	12.491

A sq.m	qD kN/sq.m	ft N/sq.mm	fb N/sq.mm	Vol. cu.m
120	2.636	-2.232	-6.446	13.4

s m	Nt kN/m	Mf kNm/m	Nf kN/m	Qf kN/m
0.00	-457.4	-0.39	-35.41	18.83
0.45	-272.3	-5.75	-12.45	6.07
0.90	-129.3	-6.73	1.92	-0.85
1.35	-37.6	-5.60	9.04	-3.65
1.80	10.8	-3.81	11.05	-4.01
2.25	29.2	-2.16	10.02	-3.24
2.70	30.4	-0.95	7.57	-2.16
3.60	15.5	0.17	2.36	-0.50
4.50	3.3	0.27	-0.84	0.12

```
TABLE  2  AP
  DIA.          RISE          RADIUS          ALPHA
   m             m              m             deg.
12.400        1.546         13.205          28.00
  h1        h2        S          b             d
  mm        mm        m          mm            mm
  75       180      2.595       250           700
  y        db       qd          qb            p
  mm       mm     kN/sq.m     kN/sq.m         m
 189      496      2.03        12.90        12.908
  A        qD        ft          fb          Vol.
sq.m    kN/sq.m   N/sq.mm     N/sq.mm        cu.m
 128     2.624    -2.236      -6.412         14.3
  s        Nt        Mf          Nf           Qf
  m       kN/m     kNm/m        kN/m         kN/m
0.00    -457.8    -0.52       -34.64        18.42
0.45    -274.5    -5.79       -12.32         6.02
0.90    -132.3    -6.79         1.71        -0.77
1.35     -40.2    -5.70         8.76        -3.57
1.80       9.2    -3.93        10.85        -3.99
2.25      28.8    -2.27         9.94        -3.28
2.70      30.9    -1.03         7.61        -2.22
3.15      24.8    -0.26         4.94        -1.26
3.60      16.6     0.14         2.53        -0.56
4.50       3.9     0.28        -0.65         0.10
```

TABLE 3 AP

DIA.		RISE		RADIUS	ALPHA
m		m		m	deg.
12.700		1.583		13.524	28.00
h1	h2	S		b	d
mm	mm	m		mm	mm
75	180	2.626		250	700
y	db	qd		qb	p
mm	mm	kN/sq.m		kN/sq.m	m
189	496	2.02		12.30	13.220
A	qD	ft		fb	Vol.
sq.m	kN/sq.m	N/sq.mm		N/sq.mm	cu.m
135	2.615	-2.240		-6.386	14.9
s	Nt	Mf		Nf	Qf
m	kN/m	kNm/m		kN/m	kN/m
0.00	-458.1	-0.62		-34.10	18.13
0.45	-276.1	-5.82		-12.24	5.99
0.90	-134.5	-6.83		1.56	-0.70
1.35	-42.1	-5.77		8.56	-3.51
1.80	8.1	-4.02		10.70	-3.97
2.25	28.5	-2.36		9.88	-3.30
2.70	31.2	-1.10		7.64	-2.27
3.15	25.5	-0.30		5.03	-1.31
4.05	9.9	0.28		0.79	-0.15
4.95	1.0	0.23		-1.34	0.17

```
TABLE  4  AP
```

DIA. m		RISE m		RADIUS m	ALPHA deg.
13.100		1.633		13.950	28.00

h1 mm	h2 mm	S m	b mm	d mm
75	180	2.667	250	700

y mm	db mm	qd kN/sq.m	qb kN/sq.m	p m
189	496	2.02	11.56	13.636

A sq.m	qD kN/sq.m	ft N/sq.mm	fb N/sq.mm	Vol. cu.m
143	2.604	−2.244	−6.352	15.8

s m	Nt kN/m	Mf kNm/m	Nf kN/m	Qf kN/m
0.00	−458.4	−0.73	−33.40	17.76
0.45	−278.2	−5.85	−12.13	5.96
0.90	−137.4	−6.88	1.37	−0.62
1.35	−44.7	−5.86	8.30	−3.43
1.80	6.5	−4.13	10.50	−3.95
2.25	28.0	−2.47	9.80	−3.33
2.70	31.6	−1.19	7.67	−2.33
3.15	26.4	−0.36	5.14	−1.38
4.05	10.8	0.27	0.96	−0.19
4.95	1.4	0.24	−1.17	0.16

TABLE 5 AP

DIA.		RISE		RADIUS	ALPHA
m		m		m	deg.
13.500		1.683		14.376	28.00
h1	h2		S	b	d
mm	mm		m	mm	mm
75	180		2.708	250	700
y	db	qd		qb	p
mm	mm	kN/sq.m		kN/sq.m	m
189	496	2.02		10.89	14.053
A	qD	ft		fb	Vol.
sq.m	kN/sq.m	N/sq.mm		N/sq.mm	cu.m
152	2.594	-2.249		-6.318	16.7
s	Nt	Mf		Nf	Qf
m	kN/m	kNm/m		kN/m	kN/m
0.00	-458.7	-0.84		-32.75	17.41
0.45	-280.2	-5.88		-12.02	5.92
0.90	-140.2	-6.93		1.19	-0.54
1.35	-47.2	-5.94		8.05	-3.35
1.80	4.9	-4.24		10.31	-3.92
2.25	27.5	-2.58		9.72	-3.35
2.70	31.9	-1.28		7.70	-2.39
3.15	27.2	-0.42		5.24	-1.45
4.05	11.7	0.26		1.12	-0.23
4.95	1.9	0.25		-1.02	0.15

TABLE 6 AP

DIA.		RISE		RADIUS		ALPHA
m		m		m		deg.
13.900		1.733		14.802		28.00

h1	h2	S	b	d
mm	mm	m	mm	mm
75	180	2.748	250	700

y	db	qd	qb	p
mm	mm	kN/sq.m	kN/sq.m	m
189	496	2.01	10.27	14.469

A	qD	ft	fb	Vol.
sq.m	kN/sq.m	N/sq.mm	N/sq.mm	cu.m
161	2.583	-2.254	-6.285	17.6

s	Nt	Mf	Nf	Qf
m	kN/m	kNm/m	kN/m	kN/m
0.00	-459.0	-0.94	-32.12	17.08
0.45	-282.1	-5.91	-11.93	5.89
0.90	-143.0	-6.97	1.01	-0.46
1.35	-49.7	-6.02	7.80	-3.28
1.80	3.3	-4.35	10.12	-3.89
2.25	26.9	-2.68	9.64	-3.37
2.70	32.2	-1.37	7.72	-2.44
3.15	28.0	-0.49	5.34	-1.51
3.60	20.3	0.02	3.09	-0.77
4.50	6.5	0.30	-0.05	0.01
5.40	-0.1	0.19	-1.35	0.17

TABLE 7 AP

DIA.		RISE		RADIUS	ALPHA
m		m		m	deg.
14.300		1.783		15.228	28.00
h1	h2	S		b	d
mm	mm	m		mm	mm
75	180	2.787		250	700
y	db	qd		qb	p
mm	mm	kN/sq.m		kN/sq.m	m
189	496	2.01		9.70	14.886
A	qD	ft		fb	Vol.
sq.m	kN/sq.m	N/sq.mm		N/sq.mm	cu.m
171	2.573	-2.258		-6.252	18.6
s	Nt	Mf		Nf	Qf
m	kN/m	kNm/m		kN/m	kN/m
0.00	-459.2	-1.04		-31.53	16.77
0.45	-284.0	-5.93		-11.84	5.85
0.90	-145.7	-7.01		0.84	-0.38
1.35	-52.2	-6.10		7.57	-3.20
1.80	1.6	-4.45		9.94	-3.86
2.25	26.2	-2.79		9.56	-3.39
2.70	32.4	-1.46		7.74	-2.49
3.15	28.7	-0.55		5.43	-1.57
3.60	21.2	-0.02		3.22	-0.83
4.50	7.2	0.30		0.09	-0.02
5.40	0.2	0.21		-1.23	0.17

TABLE 8 AP

DIA.	RISE	RADIUS	ALPHA
m	m	m	deg.
14.800	1.845	15.761	28.00

h1	h2	S	b	d
mm	mm	m	mm	mm
75	180	2.835	250	700

y	db	qd	qb	p
mm	mm	kN/sq.m	kN/sq.m	m
189	496	2.01	9.06	15.406

A	qD	ft	fb	Vol.
sq.m	kN/sq.m	N/sq.mm	N/sq.mm	cu.m
183	2.561	-2.264	-6.210	19.8

s	Nt	Mf	Nf	Qf
m	kN/m	kNm/m	kN/m	kN/m
0.00	-459.3	-1.14	-30.84	16.40
0.45	-286.2	-5.96	-11.74	5.82
0.90	-148.9	-7.05	0.63	-0.29
1.35	-55.3	-6.19	7.28	-3.10
1.80	-0.5	-4.57	9.71	-3.82
2.25	25.4	-2.92	9.45	-3.41
2.70	32.5	-1.57	7.75	-2.55
3.15	29.5	-0.63	5.53	-1.64
3.60	22.4	-0.06	3.38	-0.90
4.50	8.1	0.30	0.26	-0.05
5.40	0.6	0.22	-1.10	0.16

TABLE 9 AP

DIA.	RISE	RADIUS	ALPHA
m	m	m	deg.
15.200	1.895	16.186	28.00

h1	h2	S	b	d
mm	mm	m	mm	mm
75	180	2.873	250	700

y	db	qd	qb	p
mm	mm	kN/sq.m	kN/sq.m	m
189	496	2.00	8.59	15.822

A	qD	ft	fb	Vol.
sq.m	kN/sq.m	N/sq.mm	N/sq.mm	cu.m
193	2.552	-2.269	-6.178	20.8

s	Nt	Mf	Nf	Qf
m	kN/m	kNm/m	kN/m	kN/m
0.00	-459.5	-1.22	-30.31	16.12
0.45	-287.9	-5.97	-11.66	5.79
0.90	-151.4	-7.08	0.48	-0.22
1.35	-57.7	-6.25	7.06	-3.03
1.80	-2.2	-4.66	9.54	-3.78
2.25	24.6	-3.02	9.37	-3.42
2.70	32.6	-1.66	7.76	-2.59
3.15	30.1	-0.69	5.61	-1.70
3.60	23.2	-0.11	3.49	-0.95
4.50	8.8	0.30	0.38	-0.08
5.85	-0.7	0.16	-1.24	0.16

TABLE 10 AP

DIA.		RISE		RADIUS		ALPHA
m		m		m		deg.
15.700		1.957		16.719		28.00
h1	h2		S		b	d
mm	mm		m		mm	mm
75	180		2.920		250	700
y	db	qd		qb		p
mm	mm	kN/sq.m		kN/sq.m		m
189	496	2.00		8.05		16.343
A	qD		ft		fb	Vol.
sq.m	kN/sq.m		N/sq.mm		N/sq.mm	cu.m
206	2.541		-2.274		-6.138	22.1
s	Nt		Mf		Nf	Qf
m	kN/m		kNm/m		kN/m	kN/m
0.00	-459.5		-1.32		-29.69	15.79
0.45	-290.0		-5.99		-11.57	5.76
0.90	-154.5		-7.12		0.29	-0.13
1.35	-60.7		-6.33		6.80	-2.94
1.80	-4.3		-4.78		9.32	-3.74
2.25	23.6		-3.14		9.26	-3.42
2.70	32.6		-1.76		7.76	-2.63
3.15	30.8		-0.78		5.70	-1.76
3.60	24.2		-0.16		3.63	-1.02
4.05	16.6		0.17		1.87	-0.47
4.95	4.7		0.30		-0.37	0.07
5.85	-0.5		0.17		-1.14	0.16

```
TABLE  11  AP
  DIA.         RISE         RADIUS        ALPHA
   m            m             m           deg.
 16.100       2.007        17.145        28.00
   h1          h2           S             b            d
   mm          mm           m             mm           mm
   75          180         2.957         250          700
   y           db           qd            qb           p
   mm          mm         kN/sq.m       kN/sq.m        m
  189         496          2.00          7.65        16.759
   A           qD           ft            fb          Vol.
  sq.m       kN/sq.m      N/sq.mm       N/sq.mm       cu.m
  216         2.533       -2.278        -6.106        23.2
   s           Nt           Mf            Nf           Qf
   m          kN/m         kNm/m         kN/m         kN/m
 0.00       -459.6        -1.39        -29.22        15.54
 0.45       -291.6        -6.01        -11.50         5.73
 0.90       -156.9        -7.15          0.14        -0.07
 1.35        -63.0        -6.39          6.59        -2.86
 1.80         -6.0        -4.86          9.16        -3.70
 2.25         22.8        -3.23          9.17        -3.43
 2.70         32.5        -1.85          7.76        -2.67
 3.15         31.3        -0.84          5.76        -1.81
 3.60         25.0        -0.20          3.73        -1.07
 4.05         17.4         0.15          1.98        -0.51
 4.95          5.3         0.30         -0.27         0.05
 5.85         -0.3         0.19         -1.07         0.16
```

TABLE 12 AP

DIA.	RISE	RADIUS	ALPHA
m	m	m	deg.
16.600	2.070	17.677	28.00

h1	h2	S	b	d
mm	mm	m	mm	mm
75	180	3.003	250	700

y	db	qd	qb	p
mm	mm	kN/sq.m	kN/sq.m	m
189	496	2.00	7.20	17.280

A	qD	ft	fb	Vol.
sq.m	kN/sq.m	N/sq.mm	N/sq.mm	cu.m
230	2.523	-2.284	-6.067	24.6

s	Nt	Mf	Nf	Qf
m	kN/m	kNm/m	kN/m	kN/m
0.00	-459.6	-1.47	-28.65	15.24
0.45	-293.5	-6.02	-11.41	5.70
0.90	-159.8	-7.18	-0.03	0.01
1.35	-65.9	-6.47	6.34	-2.77
1.80	-8.1	-4.97	8.95	-3.65
2.25	21.6	-3.35	9.06	-3.42
2.70	32.4	-1.96	7.75	-2.71
3.15	31.9	-0.93	5.83	-1.87
3.60	25.9	-0.26	3.85	-1.13
4.05	18.4	0.12	2.13	-0.56
4.95	6.0	0.31	-0.15	0.03
6.30	-1.1	0.13	-1.07	0.14

TABLE 13 AP

DIA.	RISE	RADIUS	ALPHA
m	m	m	deg.
17.100	2.132	18.210	28.00

h1	h2	S	b	d
mm	mm	m	mm	mm
75	180	3.047	250	700

y	db	qd	qb	p
mm	mm	kN/sq.m	kN/sq.m	m
189	496	1.99	6.79	17.800

A	qD	ft	fb	Vol.
sq.m	kN/sq.m	N/sq.mm	N/sq.mm	cu.m
244	2.513	-2.289	-6.028	26.0

s	Nt	Mf	Nf	Qf
m	kN/m	kNm/m	kN/m	kN/m
0.00	-459.5	-1.54	-28.12	14.95
0.45	-295.3	-6.03	-11.34	5.67
0.90	-162.7	-7.20	-0.20	0.09
1.35	-68.7	-6.53	6.10	-2.69
1.80	-10.3	-5.06	8.75	-3.60
2.25	20.5	-3.46	8.95	-3.42
2.70	32.1	-2.06	7.73	-2.74
3.15	32.3	-1.01	5.90	-1.93
3.60	26.8	-0.32	3.97	-1.19
4.05	19.4	0.08	2.26	-0.62
5.40	2.8	0.28	-0.62	0.12
6.30	-0.9	0.15	-1.02	0.15

TABLE 14 AP

DIA.		RISE	RADIUS	ALPHA
m		m	m	deg.
17.600		2.194	18.742	28.00
h1	h2	S	b	d
mm	mm	m	mm	mm
75	180	3.092	250	700
y	db	qd	qb	p
mm	mm	kN/sq.m	kN/sq.m	m
189	496	1.99	6.41	18.321
A	qD	ft	fb	Vol.
sq.m	kN/sq.m	N/sq.mm	N/sq.mm	cu.m
258	2.504	-2.293	-5.990	27.4
s	Nt	Mf	Nf	Qf
m	kN/m	kNm/m	kN/m	kN/m
0.00	-459.4	-1.61	-27.61	14.68
0.45	-297.1	-6.04	-11.26	5.65
0.90	-165.4	-7.23	-0.36	0.17
1.35	-71.5	-6.59	5.87	-2.60
1.80	-12.4	-5.16	8.55	-3.54
2.25	19.3	-3.57	8.83	-3.41
2.70	31.9	-2.16	7.71	-2.77
3.15	32.7	-1.09	5.95	-1.98
3.60	27.6	-0.37	4.07	-1.25
4.05	20.4	0.05	2.39	-0.67
4.50	13.3	0.25	1.05	-0.27
5.40	3.3	0.29	-0.53	0.11
6.75	-1.4	0.10	-0.94	0.12

TABLE 15 AP

DIA.	RISE	RADIUS	ALPHA
m	m	m	deg.
18.000	2.244	19.168	28.00

h1	h2	S	b	d
mm	mm	m	mm	mm
90	200	3.334	280	800

y	db	qd	qb	p
mm	mm	kN/sq.m	kN/sq.m	m
224	573	2.37	8.82	18.737

A	qD	ft	fb	Vol.
sq.m	kN/sq.m	N/sq.mm	N/sq.mm	cu.m
270	2.929	-1.788	-4.828	33.6

s	Nt	Mf	Nf	Qf
m	kN/m	kNm/m	kN/m	kN/m
0.00	-400.3	-1.41	-25.41	13.51
0.45	-268.2	-5.62	-11.35	5.70
0.90	-158.5	-6.97	-1.57	0.74
1.35	-77.0	-6.62	4.42	-1.97
1.80	-22.7	-5.44	7.41	-3.09
2.25	9.2	-4.00	8.23	-3.21
2.70	24.5	-2.63	7.67	-2.78
3.15	28.8	-1.52	6.37	-2.14
3.60	26.7	-0.71	4.77	-1.48
4.05	21.6	-0.18	3.19	-0.91
4.50	15.6	0.13	1.82	-0.47
5.85	2.5	0.27	-0.56	0.10
6.75	-0.7	0.16	-0.98	0.13

```
TABLE  16 AP
  DIA.           RISE           RADIUS           ALPHA
   m              m                m              deg.
 18.500         2.307          19.701           28.00
   h1           h2        S         b             d
   mm           mm        m         mm            mm
   90           200      3.380      280           800
   y            db       qd         qb            p
   mm           mm    kN/sq.m    kN/sq.m          m
  224          573      2.36       8.35         19.258
   A           qD        ft         fb           Vol.
 sq.m       kN/sq.m   N/sq.mm    N/sq.mm        cu.m
  286         2.920    -1.791     -4.801         35.3
   s           Nt        Mf         Nf            Qf
   m          kN/m     kNm/m       kN/m          kN/m
 0.00        -400.4    -1.49      -24.97        13.28
 0.45        -269.6    -5.64      -11.25         5.66
 0.90        -160.7    -6.99       -1.67         0.79
 1.35         -79.3    -6.68        4.24        -1.89
 1.80         -24.6    -5.52        7.23        -3.03
 2.25           7.9    -4.09        8.11        -3.19
 2.70          24.0    -2.73        7.62        -2.80
 3.15          28.9    -1.61        6.38        -2.18
 3.60          27.2    -0.78        4.83        -1.53
 4.05          22.3    -0.22        3.29        -0.96
 4.50          16.4     0.10        1.94        -0.52
 5.85           3.0     0.28       -0.48         0.09
 7.20          -1.1     0.11       -0.95         0.12
```

TABLE 17 AP

DIA.	RISE	RADIUS	ALPHA
m	m	m	deg.
19.100	2.381	20.340	28.00

h1	h2	S	b	d
mm	mm	m	mm	mm
90	200	3.435	280	800

y	db	qd	qb	p
mm	mm	kN/sq.m	kN/sq.m	m
224	573	2.36	7.83	19.882

A	qD	ft	fb	Vol.
sq.m	kN/sq.m	N/sq.mm	N/sq.mm	cu.m
304	2.909	-1.796	-4.769	37.5

s	Nt	Mf	Nf	Qf
m	kN/m	kNm/m	kN/m	kN/m
0.00	-400.4	-1.57	-24.47	13.01
0.45	-271.2	-5.65	-11.14	5.61
0.90	-163.2	-7.02	-1.79	0.85
1.35	-82.1	-6.74	4.02	-1.81
1.80	-26.9	-5.62	7.02	-2.97
2.25	6.4	-4.21	7.96	-3.16
2.70	23.2	-2.85	7.55	-2.81
3.15	28.9	-1.72	6.39	-2.22
3.60	27.7	-0.86	4.90	-1.58
4.05	23.1	-0.28	3.40	-1.01
4.50	17.3	0.07	2.06	-0.56
4.95	11.8	0.24	0.98	-0.24
5.85	3.6	0.29	-0.38	0.08
7.20	-1.0	0.12	-0.90	0.12

TABLE 18 AP

DIA.	RISE	RADIUS	ALPHA
m	m	m	deg.
19.700	2.456	20.979	28.00

h1	h2	S	b	d
mm	mm	m	mm	mm
90	200	3.488	280	800

y	db	qd	qb	p
mm	mm	kN/sq.m	kN/sq.m	m
224	573	2.36	7.36	20.507

A	qD	ft	fb	Vol.
sq.m	kN/sq.m	N/sq.mm	N/sq.mm	cu.m
324	2.898	−1.800	−4.738	39.8

s	Nt	Mf	Nf	Qf
m	kN/m	kNm/m	kN/m	kN/m
0.00	−400.4	−1.65	−24.00	12.76
0.45	−272.8	−5.67	−11.04	5.57
0.90	−165.7	−7.04	−1.91	0.91
1.35	−84.7	−6.79	3.82	−1.73
1.80	−29.2	−5.70	6.81	−2.90
2.25	4.8	−4.32	7.81	−3.13
2.70	22.4	−2.97	7.48	−2.82
3.15	28.8	−1.82	6.39	−2.25
3.60	28.2	−0.95	4.97	−1.63
4.05	23.9	−0.34	3.50	−1.07
4.50	18.3	0.03	2.18	−0.61
4.95	12.6	0.23	1.10	−0.28
6.30	1.6	0.26	−0.63	0.12
7.65	−1.2	0.09	−0.81	0.10

TABLE 19 AP

DIA.	RISE	RADIUS	ALPHA
m	m	m	deg.
20.300	2.531	21.617	28.00

h1	h2	S	b	d
mm	mm	m	mm	mm
90	200	3.541	280	800

y	db	qd	qb	p
mm	mm	kN/sq.m	kN/sq.m	m
224	573	2.35	6.93	21.131

A	qD	ft	fb	Vol.
sq.m	kN/sq.m	N/sq.mm	N/sq.mm	cu.m
344	2.888	-1.804	-4.707	42.1

s	Nt	Mf	Nf	Qf
m	kN/m	kNm/m	kN/m	kN/m
0.00	-400.4	-1.72	-23.55	12.53
0.45	-274.2	-5.68	-10.94	5.53
0.90	-168.1	-7.06	-2.02	0.97
1.35	-87.4	-6.84	3.62	-1.64
1.80	-31.5	-5.79	6.62	-2.84
2.25	3.2	-4.42	7.66	-3.10
2.70	21.6	-3.08	7.41	-2.82
3.15	28.7	-1.92	6.39	-2.28
3.60	28.6	-1.03	5.02	-1.68
4.05	24.6	-0.41	3.60	-1.12
4.50	19.1	-0.01	2.29	-0.66
4.95	13.5	0.21	1.21	-0.32
6.30	2.0	0.27	-0.56	0.11
7.65	-1.2	0.10	-0.79	0.11

TABLE 20 AP

DIA.	RISE	RADIUS	ALPHA
m	m	m	deg.
20.900	2.606	22.256	28.00

h1	h2	S	b	d
mm	mm	m	mm	mm
90	200	3.593	280	800

y	db	qd	qb	p
mm	mm	kN/sq.m	kN/sq.m	m
224	573	2.35	6.54	21.756

A	qD	ft	fb	Vol.
sq.m	kN/sq.m	N/sq.mm	N/sq.mm	cu.m
364	2.878	-1.807	-4.677	44.4

s	Nt	Mf	Nf	Qf
m	kN/m	kNm/m	kN/m	kN/m
0.00	-400.3	-1.78	-23.13	12.30
0.45	-275.6	-5.69	-10.85	5.49
0.90	-170.4	-7.07	-2.13	1.02
1.35	-89.9	-6.89	3.43	-1.57
1.80	-33.8	-5.87	6.43	-2.78
2.25	1.6	-4.53	7.52	-3.07
2.70	20.7	-3.18	7.33	-2.82
3.15	28.5	-2.02	6.38	-2.31
3.60	28.9	-1.12	5.07	-1.72
4.05	25.3	-0.47	3.68	-1.17
4.50	20.0	-0.05	2.40	-0.71
4.95	14.4	0.18	1.32	-0.36
5.40	9.4	0.29	0.49	-0.12
6.75	0.6	0.22	-0.70	0.13
8.10	-1.3	0.07	-0.69	0.09

```
TABLE 21 AP
 DIA.        RISE       RADIUS       ALPHA
   m           m           m          deg.
21.500       2.681      22.895       28.00
  h1       h2       S         b          d
  mm       mm       m        mm         mm
  90       200     3.644     280        800
   y       db       qd        qb         p
  mm       mm    kN/sq.m   kN/sq.m      m
 224      573     2.35      6.18      22.380
   A       qD       ft        fb       Vol.
 sq.m   kN/sq.m  N/sq.mm  N/sq.mm    cu.m
 386     2.869   -1.811    -4.647      46.9
   s       Nt       Mf        Nf        Qf
   m      kN/m    kNm/m     kN/m       kN/m
```

s m	Nt kN/m	Mf kNm/m	Nf kN/m	Qf kN/m
0.00	-400.2	-1.84	-22.73	12.09
0.45	-276.9	-5.69	-10.76	5.45
0.90	-172.6	-7.08	-2.23	1.07
1.35	-92.4	-6.93	3.25	-1.49
1.80	-36.0	-5.94	6.24	-2.71
2.25	-0.1	-4.62	7.38	-3.04
2.70	19.8	-3.29	7.26	-2.82
3.15	28.2	-2.12	6.37	-2.33
3.60	29.2	-1.20	5.11	-1.76
4.05	25.9	-0.54	3.76	-1.21
4.50	20.8	-0.10	2.50	-0.75
4.95	15.2	0.16	1.43	-0.40
5.40	10.1	0.28	0.59	-0.15
6.75	0.9	0.24	-0.64	0.13
8.10	-1.3	0.08	-0.67	0.09

```
TABLE 22 AP
```

DIA.	RISE	RADIUS	ALPHA
m	m	m	deg.
22.100	2.755	23.534	28.00

h1	h2	S	b	d
mm	mm	m	mm	mm
90	200	3.695	280	800

y	db	qd	qb	p
mm	mm	kN/sq.m	kN/sq.m	m
224	573	2.34	5.85	23.005

A	qD	ft	fb	Vol.
sq.m	kN/sq.m	N/sq.mm	N/sq.mm	cu.m
408	2.860	−1.814	−4.618	49.4

s	Nt	Mf	Nf	Qf
m	kN/m	kNm/m	kN/m	kN/m
0.00	−400.1	−1.89	−22.34	11.88
0.45	−278.2	−5.70	−10.68	5.42
0.90	−174.7	−7.09	−2.32	1.12
1.35	−94.8	−6.97	3.07	−1.41
1.80	−38.2	−6.01	6.06	−2.65
2.25	−1.7	−4.72	7.24	−3.00
2.70	18.8	−3.39	7.18	−2.82
3.15	27.9	−2.22	6.35	−2.36
3.60	29.4	−1.29	5.15	−1.80
4.05	26.5	−0.60	3.84	−1.26
4.50	21.5	−0.14	2.60	−0.80
4.95	16.0	0.13	1.53	−0.44
5.40	10.9	0.27	0.69	−0.18
6.75	1.2	0.25	−0.59	0.12
8.55	−1.3	0.05	−0.57	0.07

TABLE 23 AP

DIA.		RISE		RADIUS		ALPHA
m		m		m		deg.
22.800		2.843		24.280		28.00

h1	h2		S	b	d
mm	mm		m	mm	mm
90	200		3.753	280	800

y	db	qd	qb	p
mm	mm	kN/sq.m	kN/sq.m	m
224	573	2.34	5.50	23.734

A	qD	ft	fb	Vol.
sq.m	kN/sq.m	N/sq.mm	N/sq.mm	cu.m
434	2.850	-1.818	-4.584	52.4

s	Nt	Mf	Nf	Qf
m	kN/m	kNm/m	kN/m	kN/m
0.00	-399.9	-1.95	-21.92	11.65
0.45	-279.6	-5.70	-10.58	5.38
0.90	-177.2	-7.10	-2.43	1.18
1.35	-97.6	-7.01	2.88	-1.33
1.80	-40.7	-6.09	5.86	-2.58
2.25	-3.6	-4.82	7.08	-2.96
2.70	17.7	-3.51	7.08	-2.81
3.15	27.5	-2.33	6.33	-2.38
3.60	29.6	-1.38	5.19	-1.84
4.05	27.1	-0.68	3.92	-1.31
4.50	22.4	-0.20	2.70	-0.85
4.95	16.9	0.10	1.65	-0.48
5.40	11.8	0.25	0.80	-0.22
5.85	7.4	0.31	0.16	-0.04
7.20	0.1	0.20	-0.66	0.13
8.55	-1.3	0.06	-0.57	0.08

TABLE 24 AP

DIA.	RISE	RADIUS	ALPHA
m	m	m	deg.
23.500	2.930	25.025	28.00

h1	h2	S	b	d
mm	mm	m	mm	mm
90	200	3.810	280	800

y	db	qd	qb	p
mm	mm	kN/sq.m	kN/sq.m	m
224	573	2.34	5.17	24.462

A	qD	ft	fb	Vol.
sq.m	kN/sq.m	N/sq.mm	N/sq.mm	cu.m
461	2.840	-1.822	-4.551	55.4

s	Nt	Mf	Nf	Qf
m	kN/m	kNm/m	kN/m	kN/m
0.00	-399.7	-2.00	-21.51	11.44
0.45	-280.9	-5.69	-10.49	5.34
0.90	-179.5	-7.10	-2.53	1.23
1.35	-100.3	-7.05	2.69	-1.25
1.80	-43.2	-6.16	5.66	-2.51
2.25	-5.6	-4.92	6.92	-2.92
2.70	16.5	-3.62	6.99	-2.80
3.15	27.0	-2.44	6.30	-2.39
3.60	29.7	-1.48	5.22	-1.87
4.05	27.7	-0.76	3.99	-1.35
4.50	23.2	-0.25	2.80	-0.89
4.95	17.8	0.06	1.76	-0.53
5.40	12.6	0.24	0.90	-0.25
5.85	8.2	0.30	0.26	-0.07
7.20	0.4	0.22	-0.62	0.13
9.00	-1.2	0.04	-0.47	0.06

TABLE 25 AP

DIA. m		RISE m	RADIUS m	ALPHA deg.
24.000		2.992	25.558	28.00

h1 mm	h2 mm	S m	b mm	d mm
100	250	4.230	300	1000

y mm	db mm	qd kN/sq.m	qb kN/sq.m	p m
273	717	2.68	6.12	24.983

A sq.m	qD kN/sq.m	ft N/sq.mm	fb N/sq.mm	Vol. cu.m
481	3.411	-1.268	-3.464	69.5

s m	Nt kN/m	Mf kNm/m	Nf kN/m	Qf kN/m
0.00	-358.0	-2.04	-21.34	11.35
0.45	-263.0	-5.85	-11.57	5.90
0.90	-179.7	-7.58	-4.14	2.02
1.35	-111.6	-7.87	1.12	-0.52
1.80	-59.4	-7.27	4.50	-2.00
2.25	-22.0	-6.18	6.37	-2.70
2.70	2.7	-4.92	7.08	-2.85
3.15	17.4	-3.67	6.96	-2.66
3.60	24.5	-2.55	6.30	-2.28
4.05	26.4	-1.62	5.34	-1.83
4.50	24.9	-0.90	4.25	-1.38
4.95	21.6	-0.38	3.17	-0.96
5.40	17.4	-0.02	2.18	-0.62
5.85	13.2	0.19	1.32	-0.35
7.65	1.6	0.29	-0.56	0.11
9.00	-1.0	0.13	-0.78	0.11

TABLE 26 AP

DIA. m	RISE m		RADIUS m	ALPHA deg.
24.700	3.080		26.303	28.00
h1 mm	h2 mm	S m	b mm	d mm
100	250	4.291	300	1000
y mm	db mm	qd kN/sq.m	qb kN/sq.m	p m
273	717	2.67	5.78	25.712
A sq.m	qD kN/sq.m	ft N/sq.mm	fb N/sq.mm	Vol. cu.m
509	3.398	-1.271	-3.441	73.3
s m	Nt kN/m	Mf kNm/m	Nf kN/m	Qf kN/m
0.00	-357.8	-2.10	-20.96	11.15
0.45	-264.0	-5.86	-11.45	5.84
0.90	-181.4	-7.58	-4.19	2.05
1.35	-113.7	-7.90	0.97	-0.45
1.80	-61.6	-7.33	4.32	-1.93
2.25	-24.0	-6.27	6.20	-2.64
2.70	1.3	-5.03	6.94	-2.82
3.15	16.5	-3.78	6.87	-2.66
3.60	24.1	-2.66	6.27	-2.30
4.05	26.4	-1.72	5.36	-1.86
4.50	25.3	-0.99	4.31	-1.42
4.95	22.2	-0.44	3.25	-1.01
5.40	18.1	-0.07	2.28	-0.66
5.85	13.9	0.16	1.42	-0.39
6.30	10.0	0.29	0.73	-0.18
7.65	2.0	0.30	-0.49	0.10
9.45	-1.1	0.10	-0.71	0.09

TABLE 27 AP

DIA.	RISE	RADIUS	ALPHA
m	m	m	deg.
25.500	3.179	27.155	28.00

h1	h2	S	b	d
mm	mm	m	mm	mm
100	250	4.360	300	1000

y	db	qd	qb	p
mm	mm	kN/sq.m	kN/sq.m	m
273	717	2.67	5.42	26.544

A	qD	ft	fb	Vol.
sq.m	kN/sq.m	N/sq.mm	N/sq.mm	cu.m
543	3.383	-1.275	-3.414	77.8

s	Nt	Mf	Nf	Qf
m	kN/m	kNm/m	kN/m	kN/m
0.00	-357.6	-2.17	-20.54	10.93
0.45	-265.0	-5.87	-11.31	5.78
0.90	-183.4	-7.58	-4.25	2.08
1.35	-116.2	-7.93	0.81	-0.38
1.80	-64.1	-7.39	4.12	-1.85
2.25	-26.1	-6.37	6.01	-2.58
2.70	-0.4	-5.15	6.79	-2.79
3.15	15.4	-3.91	6.78	-2.65
3.60	23.6	-2.79	6.23	-2.32
4.05	26.4	-1.84	5.37	-1.90
4.50	25.7	-1.09	4.36	-1.46
4.95	22.8	-0.52	3.34	-1.06
5.40	18.9	-0.13	2.38	-0.71
5.85	14.8	0.13	1.54	-0.43
6.30	10.8	0.27	0.83	-0.22
8.10	0.9	0.27	-0.59	0.11
9.90	-1.2	0.07	-0.63	0.08

```
TABLE 28 AP
 DIA.        RISE        RADIUS       ALPHA
  m            m           m          deg.
26.200      3.267       27.900       28.00
 h1      h2      S        b           d
 mm      mm      m        mm          mm
100     250    4.419     300         1000
  y     db      qd        qb          p
 mm     mm    kN/sq.m   kN/sq.m       m
273    717    2.66       5.14       27.273
  A      qD      ft        fb        Vol.
sq.m  kN/sq.m  N/sq.mm   N/sq.mm    cu.m
573    3.371  -1.277    -3.392       81.8
  s      Nt      Mf        Nf          Qf
  m     kN/m    kNm/m     kN/m        kN/m
0.00  -357.4  -2.22     -20.20       10.74
0.45  -265.8  -5.87     -11.20        5.73
0.90  -185.0  -7.58      -4.29        2.11
1.35  -118.2  -7.95       0.67       -0.32
1.80   -66.2  -7.44       3.95       -1.78
2.25   -28.0  -6.45       5.84       -2.53
2.70    -1.9  -5.24       6.66       -2.75
3.15    14.4  -4.02       6.69       -2.64
3.60    23.1  -2.90       6.19       -2.33
4.05    26.4  -1.94       5.37       -1.92
4.50    26.0  -1.17       4.41       -1.50
4.95    23.4  -0.59       3.41       -1.10
5.40    19.6  -0.18       2.46       -0.75
5.85    15.5   0.09       1.63       -0.47
6.30    11.5   0.25       0.92       -0.25
6.75     8.0   0.33       0.37       -0.09
8.10     1.2   0.28      -0.54        0.11
9.90    -1.1   0.08      -0.62        0.08
```

```
TABLE  29 AP
  DIA.        RISE        RADIUS        ALPHA
   m           m            m           deg.
 27.000       3.366       28.752        28.00
   h1         h2       S        b         d
   mm         mm       m        mm        mm
  100        250      4.486     300      1000
   y         db       qd        qb        p
   mm        mm    kN/sq.m    kN/sq.m     m
  273       717      2.66       4.84     28.106
   A         qD       ft        fb       Vol.
  sq.m    kN/sq.m  N/sq.mm   N/sq.mm    cu.m
  608      3.358    -1.280    -3.366     86.5
   s         Nt       Mf        Nf        Qf
   m        kN/m     kNm/m     kN/m      kN/m
  0.00    -357.1    -2.28    -19.82     10.54
  0.45    -266.7    -5.87    -11.08      5.67
  0.90    -186.8    -7.58     -4.34      2.14
  1.35    -120.5    -7.97      0.52     -0.25
  1.80     -68.5    -7.50      3.76     -1.71
  2.25     -30.1    -6.53      5.66     -2.46
  2.70      -3.6    -5.35      6.51     -2.71
  3.15      13.3    -4.14      6.59     -2.63
  3.60      22.5    -3.02      6.14     -2.34
  4.05      26.3    -2.05      5.38     -1.95
  4.50      26.3    -1.27      4.45     -1.53
  4.95      23.9    -0.67      3.48     -1.14
  5.40      20.3    -0.24      2.55     -0.79
  5.85      16.3     0.05      1.73     -0.51
  6.30      12.3     0.23      1.03     -0.28
  6.75       8.7     0.32      0.46     -0.12
  8.55       0.4     0.24     -0.60      0.12
 10.35      -1.1     0.06     -0.54      0.07
```

TABLE 30 AP

DIA. m	RISE m	RADIUS m	ALPHA deg.
27.800	3.466	29.604	28.00

h1 mm	h2 mm	S m	b mm	d mm
100	250	4.552	300	1000

y mm	db mm	qd kN/sq.m	qb kN/sq.m	p m
273	717	2.66	4.56	28.938

A sq.m	qD kN/sq.m	ft N/sq.mm	fb N/sq.mm	Vol. cu.m
645	3.345	-1.283	-3.341	91.4

s m	Nt kN/m	Mf kNm/m	Nf kN/m	Qf kN/m
0.00	-356.7	-2.33	-19.47	10.35
0.45	-267.6	-5.87	-10.96	5.62
0.90	-188.5	-7.57	-4.39	2.17
1.35	-122.7	-7.98	0.38	-0.18
1.80	-70.8	-7.54	3.58	-1.64
2.25	-32.2	-6.61	5.49	-2.40
2.70	-5.3	-5.46	6.37	-2.67
3.15	12.1	-4.26	6.49	-2.61
3.60	21.9	-3.14	6.09	-2.34
4.05	26.1	-2.16	5.37	-1.97
4.50	26.5	-1.37	4.48	-1.57
4.95	24.4	-0.75	3.55	-1.18
5.40	21.0	-0.30	2.64	-0.83
5.85	17.0	0.01	1.82	-0.55
6.30	13.1	0.20	1.12	-0.32
6.75	9.5	0.31	0.55	-0.15
7.20	6.4	0.34	0.11	-0.03
9.00	-0.3	0.20	-0.62	0.12
10.80	-1.1	0.04	-0.46	0.06

TABLE 31 AP

DIA.	RISE	RADIUS	ALPHA
m	m	m	deg.
28.700	3.578	30.563	28.00

h1	h2	S	b	d
mm	mm	m	mm	mm
100	250	4.625	300	1000

y	db	qd	qb	p
mm	mm	kN/sq.m	kN/sq.m	m
273	717	2.65	4.28	29.875

A	qD	ft	fb	Vol.
sq.m	kN/sq.m	N/sq.mm	N/sq.mm	cu.m
687	3.331	-1.285	-3.313	97.0

s	Nt	Mf	Nf	Qf
m	kN/m	kNm/m	kN/m	kN/m
0.00	-356.3	-2.38	-19.08	10.15
0.45	-268.5	-5.86	-10.83	5.56
0.90	-190.4	-7.56	-4.44	2.20
1.35	-125.1	-8.00	0.23	-0.11
1.80	-73.3	-7.59	3.39	-1.55
2.25	-34.5	-6.70	5.29	-2.33
2.70	-7.1	-5.57	6.21	-2.63
3.15	10.7	-4.38	6.38	-2.59
3.60	21.1	-3.27	6.04	-2.35
4.05	25.8	-2.29	5.36	-2.00
4.50	26.6	-1.47	4.52	-1.61
4.95	24.9	-0.84	3.61	-1.22
5.40	21.7	-0.37	2.73	-0.88
5.85	17.8	-0.04	1.92	-0.59
6.30	13.9	0.17	1.22	-0.36
6.75	10.2	0.29	0.65	-0.18
7.20	7.1	0.34	0.20	-0.05
9.00	-0.0	0.22	-0.58	0.11
10.80	-1.1	0.05	-0.46	0.06

```
TABLE 32 AP
```

DIA.		RISE		RADIUS		ALPHA
m		m		m		deg.
29.500		3.678		31.415		28.00

h1	h2	S	b	d
mm	mm	m	mm	mm
100	250	4.689	300	1000

y	db	qd	qb	p
mm	mm	kN/sq.m	kN/sq.m	m
273	717	2.65	4.05	30.708

A	qD	ft	fb	Vol.
sq.m	kN/sq.m	N/sq.mm	N/sq.mm	cu.m
726	3.319	-1.288	-3.289	102.1

s	Nt	Mf	Nf	Qf
m	kN/m	kNm/m	kN/m	kN/m
0.00	-355.9	-2.42	-18.76	9.98
0.45	-269.2	-5.86	-10.73	5.51
0.90	-192.0	-7.55	-4.48	2.22
1.35	-127.1	-8.01	0.10	-0.05
1.80	-75.5	-7.63	3.23	-1.48
2.25	-36.5	-6.77	5.13	-2.27
2.70	-8.8	-5.66	6.07	-2.59
3.15	9.5	-4.49	6.28	-2.57
3.60	20.3	-3.38	5.98	-2.35
4.05	25.5	-2.39	5.35	-2.01
4.50	26.7	-1.57	4.54	-1.64
4.95	25.3	-0.92	3.67	-1.26
5.40	22.3	-0.43	2.80	-0.92
5.85	18.5	-0.09	2.01	-0.63
6.30	14.6	0.14	1.31	-0.39
6.75	10.9	0.27	0.73	-0.21
7.20	7.7	0.33	0.28	-0.07
7.65	5.0	0.35	-0.07	0.02
9.45	-0.5	0.18	-0.58	0.11
11.25	-1.1	0.04	-0.39	0.05

TABLE 33 AP

DIA. m	RISE m	RADIUS m	ALPHA deg.
30.000	3.740	31.947	28.00

h1 mm	h2 mm	S m	b mm	d mm
130	290	5.180	300	1200

y mm	db mm	qd kN/sq.m	qb kN/sq.m	p m
345	872	3.40	6.62	31.229

A sq.m	qD kN/sq.m	ft N/sq.mm	fb N/sq.mm	Vol. cu.m
751	4.169	-1.013	-2.786	132.7

s m	Nt kN/m	Mf kNm/m	Nf kN/m	Qf kN/m
0.00	-333.7	-3.02	-18.62	9.90
0.45	-258.3	-6.50	-11.16	5.74
0.90	-190.4	-8.34	-5.22	2.59
1.35	-132.2	-8.96	-0.69	0.33
1.80	-84.5	-8.74	2.57	-1.18
2.25	-47.1	-7.99	4.72	-2.10
2.70	-19.1	-6.92	5.97	-2.55
3.15	0.7	-5.74	6.51	-2.68
3.60	13.7	-4.55	6.50	-2.57
4.05	21.3	-3.45	6.11	-2.32
4.50	24.8	-2.48	5.48	-1.99
4.95	25.4	-1.66	4.70	-1.63
5.40	24.0	-1.01	3.86	-1.28
5.85	21.4	-0.51	3.04	-0.96
6.30	18.2	-0.14	2.26	-0.68
6.75	14.8	0.11	1.57	-0.45
7.20	11.5	0.27	0.98	-0.26
7.65	8.5	0.36	0.48	-0.12
9.45	0.9	0.30	-0.56	0.11
11.70	-1.1	0.08	-0.58	0.07

TABLE 34 AP

DIA.	RISE	RADIUS	ALPHA
m	m	m	deg.
30.900	3.853	32.905	28.00

h1	h2	S	b	d
mm	mm	m	mm	mm
130	290	5.257	300	1200

y	db	qd	qb	p
mm	mm	kN/sq.m	kN/sq.m	m
345	872	3.39	6.24	32.165

A	qD	ft	fb	Vol.
sq.m	kN/sq.m	N/sq.mm	N/sq.mm	cu.m
797	4.155	-1.015	-2.764	140.3

s	Nt	Mf	Nf	Qf
m	kN/m	kNm/m	kN/m	kN/m
0.00	-333.4	-3.07	-18.28	9.72
0.45	-259.0	-6.50	-11.04	5.68
0.90	-191.8	-8.33	-5.24	2.60
1.35	-134.1	-8.97	-0.81	0.39
1.80	-86.7	-8.78	2.40	-1.11
2.25	-49.2	-8.06	4.54	-2.03
2.70	-21.0	-7.02	5.80	-2.50
3.15	-0.9	-5.86	6.37	-2.64
3.60	12.6	-4.68	6.40	-2.55
4.05	20.7	-3.58	6.06	-2.32
4.50	24.6	-2.60	5.46	-2.01
4.95	25.5	-1.78	4.72	-1.66
5.40	24.4	-1.11	3.92	-1.32
5.85	22.0	-0.59	3.11	-1.00
6.30	18.8	-0.20	2.35	-0.72
6.75	15.5	0.07	1.67	-0.49
7.20	12.2	0.24	1.07	-0.30
7.65	9.2	0.34	0.58	-0.15
9.90	0.3	0.27	-0.60	0.11
11.70	-1.1	0.09	-0.57	0.08

TABLE 35 AP

DIA.	RISE	RADIUS	ALPHA
m	m	m	deg.
31.800	3.965	33.864	28.00

h1	h2	S	b	d
mm	mm	m	mm	mm
130	290	5.333	300	1200

y	db	qd	qb	p
mm	mm	kN/sq.m	kN/sq.m	m
345	872	3.39	5.89	33.102

A	qD	ft	fb	Vol.
sq.m	kN/sq.m	N/sq.mm	N/sq.mm	cu.m
844	4.141	-1.017	-2.742	148.0

s	Nt	Mf	Nf	Qf
m	kN/m	kNm/m	kN/m	kN/m
0.00	-333.0	-3.12	-17.96	9.55
0.45	-259.6	-6.49	-10.91	5.62
0.90	-193.2	-8.31	-5.26	2.62
1.35	-136.0	-8.97	-0.92	0.44
1.80	-88.8	-8.81	2.23	-1.04
2.25	-51.3	-8.12	4.36	-1.96
2.70	-22.9	-7.11	5.64	-2.44
3.15	-2.4	-5.97	6.23	-2.60
3.60	11.5	-4.81	6.30	-2.54
4.05	20.0	-3.71	6.00	-2.32
4.50	24.3	-2.73	5.44	-2.02
4.95	25.5	-1.89	4.74	-1.69
5.40	24.7	-1.21	3.96	-1.35
5.85	22.5	-0.67	3.18	-1.04
6.30	19.5	-0.27	2.43	-0.76
6.75	16.2	0.02	1.76	-0.52
7.20	12.9	0.21	1.16	-0.33
7.65	9.9	0.32	0.67	-0.18
8.10	7.2	0.38	0.26	-0.07
9.90	0.6	0.29	-0.55	0.11
12.15	-1.1	0.07	-0.51	0.07

TABLE 36 AP

DIA.	RISE	RADIUS	ALPHA
m	m	m	deg.
32.800	4.090	34.929	28.00

h1	h2	S	b	d
mm	mm	m	mm	mm
130	290	5.417	300	1200

y	db	qd	qb	p
mm	mm	kN/sq.m	kN/sq.m	m
345	872	3.39	5.54	34.143

A	qD	ft	fb	Vol.
sq.m	kN/sq.m	N/sq.mm	N/sq.mm	cu.m
898	4.126	-1.019	-2.718	157.0

s	Nt	Mf	Nf	Qf
m	kN/m	kNm/m	kN/m	kN/m
0.00	-332.6	-3.16	-17.62	9.37
0.45	-260.2	-6.48	-10.78	5.56
0.90	-194.7	-8.29	-5.28	2.64
1.35	-138.1	-8.97	-1.04	0.50
1.80	-91.1	-8.84	2.06	-0.96
2.25	-53.6	-8.18	4.17	-1.89
2.70	-24.9	-7.21	5.46	-2.38
3.15	-4.1	-6.09	6.08	-2.56
3.60	10.2	-4.94	6.19	-2.51
4.05	19.1	-3.85	5.93	-2.32
4.50	23.9	-2.86	5.42	-2.04
4.95	25.5	-2.02	4.75	-1.72
5.40	25.0	-1.32	4.01	-1.39
5.85	23.0	-0.76	3.25	-1.08
6.30	20.2	-0.34	2.52	-0.80
6.75	16.9	-0.03	1.85	-0.56
7.20	13.7	0.17	1.26	-0.37
7.65	10.6	0.30	0.76	-0.21
8.10	7.9	0.37	0.35	-0.09
10.35	0.0	0.25	-0.57	0.11
12.60	-1.1	0.05	-0.44	0.06

TABLE 37 AP

DIA.		RISE		RADIUS		ALPHA
m		m		m		deg.
33.800		4.214		35.994		28.00

h1	h2		S	b		d
mm	mm		m	mm		mm
130	290		5.499	300		1200

y	db		qd	qb		p
mm	mm		kN/sq.m	kN/sq.m		m
345	872		3.38	5.22		35.184

A	qD		ft	fb		Vol.
sq.m	kN/sq.m		N/sq.mm	N/sq.mm		cu.m
953	4.112		−1.021	−2.695		166.1

s	Nt	Mf	Nf	Qf
m	kN/m	kNm/m	kN/m	kN/m
0.00	−332.1	−3.20	−17.30	9.20
0.45	−260.8	−6.47	−10.66	5.50
0.90	−196.1	−8.27	−5.30	2.65
1.35	−140.0	−8.97	−1.15	0.56
1.80	−93.3	−8.87	1.90	−0.89
2.25	−55.8	−8.24	3.99	−1.81
2.70	−27.0	−7.30	5.29	−2.32
3.15	−5.8	−6.20	5.94	−2.52
3.60	8.9	−5.07	6.08	−2.49
4.05	18.3	−3.98	5.87	−2.32
4.50	23.4	−3.00	5.39	−2.05
4.95	25.5	−2.14	4.76	−1.74
5.40	25.2	−1.43	4.05	−1.43
5.85	23.4	−0.86	3.31	−1.12
6.30	20.8	−0.41	2.60	−0.84
6.75	17.7	−0.09	1.94	−0.60
7.20	14.4	0.13	1.35	−0.40
7.65	11.3	0.28	0.85	−0.24
8.10	8.5	0.36	0.43	−0.12
8.55	6.1	0.39	0.10	−0.03
10.80	−0.4	0.22	−0.56	0.11
13.05	−1.0	0.04	−0.38	0.05

```
TABLE  38 AP
```

DIA. m	RISE m		RADIUS m	ALPHA deg.
34.800	4.339		37.059	28.00

h1 mm	h2 mm	S m	b mm	d mm
130	290	5.579	300	1200

y mm	db mm	qd kN/sq.m	qb kN/sq.m	p m
345	872	3.38	4.92	36.225

A sq.m	qD kN/sq.m	ft N/sq.mm	fb N/sq.mm	Vol. cu.m
1010	4.099	-1.023	-2.672	175.5

s m	Nt kN/m	Mf kNm/m	Nf kN/m	Qf kN/m
0.00	-331.6	-3.23	-16.99	9.04
0.45	-261.3	-6.45	-10.54	5.44
0.90	-197.4	-8.25	-5.32	2.66
1.35	-141.9	-8.96	-1.26	0.61
1.80	-95.4	-8.89	1.74	-0.82
2.25	-58.0	-8.30	3.82	-1.74
2.70	-29.0	-7.38	5.12	-2.26
3.15	-7.5	-6.31	5.79	-2.47
3.60	7.6	-5.19	5.98	-2.47
4.05	17.4	-4.11	5.79	-2.31
4.50	23.0	-3.12	5.36	-2.06
4.95	25.3	-2.26	4.76	-1.77
5.40	25.4	-1.54	4.08	-1.46
5.85	23.9	-0.95	3.37	-1.16
6.30	21.3	-0.49	2.67	-0.88
6.75	18.3	-0.15	2.02	-0.64
7.20	15.2	0.09	1.44	-0.44
7.65	12.1	0.25	0.94	-0.27
8.10	9.2	0.34	0.52	-0.14
8.55	6.7	0.38	0.18	-0.05
9.00	4.6	0.39	-0.09	0.02
11.25	-0.7	0.19	-0.55	0.10
13.50	-1.0	0.03	-0.33	0.04

TABLE 39 AP

DIA.	RISE	RADIUS	ALPHA
m	m	m	deg.
35.000	4.364	37.272	28.00

h1	h2	S	b	d
mm	mm	m	mm	mm
130	290	5.595	300	1200

y	db	qd	qb	p
mm	mm	kN/sq.m	kN/sq.m	m
345	872	3.38	4.87	36.433

A	qD	ft	fb	Vol.
sq.m	kN/sq.m	N/sq.mm	N/sq.mm	cu.m
1022	4.096	-1.023	-2.668	177.4

s	Nt	Mf	Nf	Qf
m	kN/m	kNm/m	kN/m	kN/m
0.00	-331.5	-3.24	-16.93	9.00
0.45	-261.4	-6.45	-10.51	5.43
0.90	-197.7	-8.24	-5.32	2.67
1.35	-142.3	-8.96	-1.28	0.62
1.80	-95.8	-8.90	1.71	-0.81
2.25	-58.4	-8.31	3.78	-1.73
2.70	-29.4	-7.40	5.09	-2.25
3.15	-7.8	-6.33	5.76	-2.47
3.60	7.4	-5.21	5.95	-2.46
4.05	17.2	-4.14	5.78	-2.31
4.50	22.8	-3.15	5.35	-2.06
4.95	25.3	-2.29	4.76	-1.77
5.40	25.4	-1.56	4.08	-1.46
5.85	23.9	-0.97	3.38	-1.16
6.30	21.5	-0.51	2.69	-0.89
6.75	18.5	-0.16	2.04	-0.65
7.20	15.3	0.08	1.46	-0.44
7.65	12.2	0.24	0.95	-0.28
8.10	9.4	0.34	0.53	-0.15
8.55	6.8	0.38	0.19	-0.05
9.00	4.7	0.39	-0.07	0.02
11.25	-0.6	0.19	-0.54	0.10
13.50	-1.0	0.03	-0.33	0.04

TABLE 40 AP

DIA.	RISE	RADIUS	ALPHA
m	m	m	deg.
36.100	4.501	38.443	28.00

h1	h2	S	b	d
mm	mm	m	mm	mm
130	290	5.683	300	1200

y	db	qd	qb	p
mm	mm	kN/sq.m	kN/sq.m	m
345	872	3.37	4.57	37.578

A	qD	ft	fb	Vol.
sq.m	kN/sq.m	N/sq.mm	N/sq.mm	cu.m
1087	4.082	-1.025	-2.644	188.1

s	Nt	Mf	Nf	Qf
m	kN/m	kNm/m	kN/m	kN/m
0.00	-330.9	-3.27	-16.61	8.83
0.45	-261.9	-6.43	-10.39	5.37
0.90	-199.0	-8.21	-5.34	2.68
1.35	-144.2	-8.95	-1.39	0.68
1.80	-98.1	-8.91	1.54	-0.73
2.25	-60.7	-8.36	3.60	-1.65
2.70	-31.5	-7.48	4.91	-2.18
3.15	-9.6	-6.44	5.61	-2.42
3.60	5.9	-5.34	5.84	-2.43
4.05	16.2	-4.27	5.70	-2.30
4.50	22.2	-3.29	5.31	-2.07
4.95	25.1	-2.42	4.76	-1.79
5.40	25.5	-1.68	4.11	-1.49
5.85	24.3	-1.07	3.43	-1.20
6.30	22.0	-0.59	2.76	-0.93
6.75	19.2	-0.23	2.12	-0.69
7.20	16.1	0.03	1.55	-0.48
7.65	13.0	0.21	1.05	-0.31
8.10	10.1	0.32	0.62	-0.18
8.55	7.5	0.37	0.27	-0.07
9.00	5.3	0.39	0.00	0.00
11.70	-0.8	0.17	-0.51	0.10
13.95	-0.9	0.02	-0.28	0.04

Tables 1BP to 40BP   For Type B Domes

```
TABLE  1  BP
```

DIA.	RISE	RADIUS	ALPHA
m	m	m	deg.
12.000	1.496	12.779	28.00

h1	h2	S	b	d
mm	mm	m	mm	mm
75	180	2.553	250	700

y	db	qd	qb	p
mm	mm	kN/sq.m	kN/sq.m	m
-41	266	2.03	13.78	12.491

A	qD	ft	fb	Vol.
sq.m	kN/sq.m	N/sq.mm	N/sq.mm	cu.m
120	2.636	-1.915	-5.127	13.4

s	Nt	Mf	Nf	Qf
m	kN/m	kNm/m	kN/m	kN/m
0.00	-535.4	8.11	-56.48	30.04
0.45	-369.7	-1.58	-28.81	14.04
0.90	-212.6	-5.41	-8.73	3.88
1.35	-95.5	-5.81	3.50	-1.41
1.80	-22.8	-4.64	9.29	-3.37
2.25	13.9	-3.06	10.63	-3.44
2.70	26.5	-1.67	9.37	-2.67
3.60	19.6	-0.10	4.34	-0.91
4.50	6.4	0.26	0.21	-0.03

TABLE 2 BP

DIA.		RISE	RADIUS	ALPHA
m		m	m	deg.
12.400		1.546	13.205	28.00
h1	h2	S	b	d
mm	mm	m	mm	mm
75	180	2.595	250	700
y	db	qd	qb	p
mm	mm	kN/sq.m	kN/sq.m	m
-41	266	2.03	12.90	12.908
A	qD	ft	fb	Vol.
sq.m	kN/sq.m	N/sq.mm	N/sq.mm	cu.m
128	2.624	-1.906	-5.107	14.3
s	Nt	Mf	Nf	Qf
m	kN/m	kNm/m	kN/m	kN/m
0.00	-534.2	7.85	-55.02	29.26
0.45	-370.2	-1.63	-28.22	13.79
0.90	-214.9	-5.41	-8.74	3.91
1.35	-98.4	-5.85	3.20	-1.30
1.80	-25.2	-4.72	8.95	-3.29
2.25	12.6	-3.16	10.38	-3.42
2.70	26.3	-1.77	9.26	-2.71
3.15	26.4	-0.76	6.98	-1.78
3.60	20.4	-0.14	4.45	-0.98
4.50	7.1	0.25	0.41	-0.06

TABLE 3 BP

DIA.		RISE		RADIUS		ALPHA
m		m		m		deg.
12.700		1.583		13.524		28.00
h1	h2		S		b	d
mm	mm		m		mm	mm
75	180		2.626		250	700
y	db		qd		qb	p
mm	mm		kN/sq.m		kN/sq.m	m
-41	266		2.02		12.30	13.220
A	qD		ft		fb	Vol.
sq.m	kN/sq.m		N/sq.mm		N/sq.mm	cu.m
135	2.615		-1.900		-5.092	14.9

s	Nt	Mf	Nf	Qf
m	kN/m	kNm/m	kN/m	kN/m
0.00	-533.3	7.66	-53.97	28.70
0.45	-370.5	-1.66	-27.80	13.62
0.90	-216.6	-5.41	-8.75	3.93
1.35	-100.5	-5.87	2.98	-1.22
1.80	-27.0	-4.78	8.70	-3.23
2.25	11.6	-3.24	10.20	-3.41
2.70	26.0	-1.84	9.18	-2.73
3.15	26.7	-0.81	6.98	-1.83
4.05	13.9	0.14	2.30	-0.44
4.95	3.2	0.24	-0.73	0.09

```
TABLE  4  BP
 DIA.          RISE          RADIUS          ALPHA
  m             m              m              deg.
13.100       1.633        13.950           28.00
 h1        h2        S          b             d
 mm        mm        m          mm            mm
 75       180       2.667      250           700
  y        db        qd         qb            p
 mm        mm     kN/sq.m    kN/sq.m          m
-41       266      2.02       11.56        13.636
  A        qD        ft         fb          Vol.
sq.m    kN/sq.m   N/sq.mm    N/sq.mm        cu.m
143      2.604    -1.894     -5.071         15.8
  s        Nt        Mf         Nf            Qf
  m       kN/m     kNm/m       kN/m          kN/m
0.00    -532.1      7.42      -52.64        28.00
0.45    -370.9     -1.69      -27.26        13.39
0.90    -218.7     -5.41       -8.76         3.96
1.35    -103.3     -5.91        2.72        -1.12
1.80     -29.3     -4.86        8.39        -3.16
2.25      10.2     -3.34        9.97        -3.39
2.70      25.7     -1.94        9.07        -2.76
3.15      27.1     -0.89        6.99        -1.88
4.05      14.8      0.12        2.44        -0.49
4.95       3.8      0.25       -0.53         0.07
```

TABLE 5 BP

DIA.	RISE	RADIUS	ALPHA
m	m	m	deg.
13.500	1.683	14.376	28.00

h1	h2	S	b	d
mm	mm	m	mm	mm
75	180	2.708	250	700

y	db	qd	qb	p
mm	mm	kN/sq.m	kN/sq.m	m
-41	266	2.02	10.89	14.053

A	qD	ft	fb	Vol.
sq.m	kN/sq.m	N/sq.mm	N/sq.mm	cu.m
152	2.594	-1.888	-5.051	16.7

s	Nt	Mf	Nf	Qf
m	kN/m	kNm/m	kN/m	kN/m
0.00	-530.8	7.19	-51.39	27.33
0.45	-371.3	-1.73	-26.76	13.17
0.90	-220.7	-5.41	-8.76	3.98
1.35	-106.0	-5.94	2.47	-1.03
1.80	-31.6	-4.93	8.10	-3.08
2.25	8.8	-3.43	9.74	-3.36
2.70	25.2	-2.03	8.96	-2.78
3.15	27.4	-0.97	6.99	-1.93
4.05	15.6	0.09	2.57	-0.54
4.95	4.4	0.26	-0.36	0.05

```
TABLE 6 BP
  DIA.          RISE         RADIUS        ALPHA
   m             m             m            deg.
 13.900        1.733        14.802         28.00
   h1      h2        S          b            d
   mm      mm        m          mm           mm
   75     180      2.748       250          700
   y       db       qd         qb            p
   mm      mm    kN/sq.m    kN/sq.m          m
  -41     266     2.01       10.27        14.469
   A       qD       ft         fb          Vol.
 sq.m   kN/sq.m  N/sq.mm    N/sq.mm        cu.m
  161    2.583   -1.883     -5.030         17.6
   s       Nt       Mf         Nf            Qf
   m      kN/m    kNm/m      kN/m          kN/m
 0.00   -529.5    6.98      -50.20         26.69
 0.45   -371.6   -1.77      -26.27         12.96
 0.90   -222.7   -5.41       -8.76          4.00
 1.35   -108.6   -5.97        2.23         -0.94
 1.80    -33.9   -5.00        7.82         -3.01
 2.25      7.3   -3.52        9.53         -3.34
 2.70     24.7   -2.12        8.86         -2.80
 3.15     27.6   -1.04        6.99         -1.98
 3.60     23.3   -0.33        4.77         -1.20
 4.50     10.0    0.23        1.02         -0.19
 5.40      1.6    0.22       -1.02          0.13
```

TABLE 7 BP

DIA.	RISE	RADIUS	ALPHA
m	m	m	deg.
14.300	1.783	15.228	28.00

h1	h2	S	b	d
mm	mm	m	mm	mm
75	180	2.787	250	700

y	db	qd	qb	p
mm	mm	kN/sq.m	kN/sq.m	m
-41	266	2.01	9.70	14.886

A	qD	ft	fb	Vol.
sq.m	kN/sq.m	N/sq.mm	N/sq.mm	cu.m
171	2.573	-1.879	-5.009	18.6

s	Nt	Mf	Nf	Qf
m	kN/m	kNm/m	kN/m	kN/m
0.00	-528.3	6.77	-49.07	26.09
0.45	-371.9	-1.80	-25.81	12.76
0.90	-224.6	-5.41	-8.76	4.01
1.35	-111.1	-5.99	2.01	-0.85
1.80	-36.1	-5.06	7.55	-2.93
2.25	5.9	-3.61	9.33	-3.31
2.70	24.2	-2.21	8.75	-2.82
3.15	27.8	-1.12	6.99	-2.02
3.60	23.9	-0.39	4.84	-1.25
4.50	10.7	0.22	1.16	-0.23
5.40	2.0	0.23	-0.87	0.12

```
TABLE 8 BP
 DIA.          RISE          RADIUS        ALPHA
  m             m              m           deg.
14.800        1.845         15.761        28.00
 h1       h2          S          b           d
 mm       mm          m          mm          mm
 75       180        2.835      250         700
  y       db         qd         qb           p
 mm       mm      kN/sq.m    kN/sq.m         m
-41       266       2.01       9.06        15.406
  A       qD          ft         fb        Vol.
sq.m   kN/sq.m    N/sq.mm    N/sq.mm       cu.m
183      2.561     -1.874     -4.982        19.8
  s       Nt         Mf         Nf          Qf
  m      kN/m      kNm/m       kN/m         kN/m
0.00   -526.7       6.53      -47.73       25.38
0.45   -372.2      -1.84      -25.27       12.52
0.90   -226.8      -5.41       -8.76        4.03
1.35   -114.1      -6.02        1.74       -0.74
1.80    -38.9      -5.14        7.23       -2.84
2.25      4.1      -3.71        9.08       -3.27
2.70     23.5      -2.32        8.63       -2.83
3.15     28.0      -1.21        6.98       -2.07
3.60     24.6      -0.45        4.92       -1.31
4.50     11.7       0.20        1.32       -0.27
5.40      2.5       0.24       -0.71        0.10
```

TABLE 9 BP

DIA.	RISE	RADIUS	ALPHA
m	m	m	deg.
15.200	1.895	16.186	28.00

h1	h2	S	b	d
mm	mm	m	mm	mm
75	180	2.873	250	700

y	db	qd	qb	p
mm	mm	kN/sq.m	kN/sq.m	m
-41	266	2.00	8.59	15.822

A	qD	ft	fb	Vol.
sq.m	kN/sq.m	N/sq.mm	N/sq.mm	cu.m
193	2.552	-1.872	-4.961	20.8

s	Nt	Mf	Nf	Qf
m	kN/m	kNm/m	kN/m	kN/m
0.00	-525.4	6.34	-46.72	24.85
0.45	-372.4	-1.87	-24.85	12.34
0.90	-228.5	-5.40	-8.75	4.05
1.35	-116.5	-6.05	1.55	-0.66
1.80	-41.1	-5.19	6.99	-2.77
2.25	2.6	-3.80	8.89	-3.24
2.70	22.8	-2.41	8.53	-2.84
3.15	28.1	-1.29	6.97	-2.11
3.60	25.2	-0.51	4.98	-1.36
4.50	12.4	0.19	1.44	-0.31
5.85	0.6	0.19	-1.08	0.14

```
TABLE  10  BP
 DIA.          RISE         RADIUS        ALPHA
  m             m             m            deg.
15.700        1.957        16.719        28.00
 h1        h2        S         b          d
 mm        mm        m         mm         mm
 75        180      2.920      250        700
 y         db        qd        qb         p
 mm        mm     kN/sq.m   kN/sq.m       m
-41        266      2.00      8.05      16.343
 A         qD        ft        fb        Vol.
sq.m    kN/sq.m   N/sq.mm   N/sq.mm     cu.m
206       2.541    -1.869    -4.934      22.1
 s         Nt        Mf        Nf         Qf
 m        kN/m     kNm/m      kN/m       kN/m
0.00     -523.8     6.12     -45.52      24.21
0.45     -372.6    -1.90     -24.36      12.12
0.90     -230.6    -5.40      -8.74       4.06
1.35     -119.4    -6.07       1.31      -0.57
1.80      -43.7    -5.26       6.70      -2.69
2.25        0.8    -3.89       8.66      -3.20
2.70       22.0    -2.51       8.40      -2.85
3.15       28.1    -1.38       6.95      -2.15
3.60       25.8    -0.58       5.05      -1.42
4.05       19.9    -0.08       3.17      -0.80
4.95        7.7     0.26       0.37      -0.07
5.85        0.9     0.20      -0.95       0.13
```

```
TABLE  11  BP
```

DIA.	RISE	RADIUS	ALPHA
m	m	m	deg.
16.100	2.007	17.145	28.00

h1	h2	S	b	d
mm	mm	m	mm	mm
75	180	2.957	250	700

y	db	qd	qb	p
mm	mm	kN/sq.m	kN/sq.m	m
-41	266	2.00	7.65	16.759

A	qD	ft	fb	Vol.
sq.m	kN/sq.m	N/sq.mm	N/sq.mm	cu.m
216	2.533	-1.867	-4.913	23.2

s	Nt	Mf	Nf	Qf
m	kN/m	kNm/m	kN/m	kN/m
0.00	-522.5	5.95	-44.61	23.72
0.45	-372.8	-1.93	-23.98	11.96
0.90	-232.1	-5.39	-8.73	4.07
1.35	-121.6	-6.09	1.14	-0.49
1.80	-45.8	-5.31	6.48	-2.62
2.25	-0.7	-3.97	8.48	-3.17
2.70	21.3	-2.59	8.30	-2.86
3.15	28.0	-1.45	6.93	-2.18
3.60	26.2	-0.63	5.09	-1.46
4.05	20.5	-0.12	3.26	-0.84
4.95	8.3	0.25	0.48	-0.10
5.85	1.2	0.21	-0.85	0.13

```
TABLE  12  BP
   DIA.          RISE            RADIUS          ALPHA
    m              m                m             deg.
 16.600         2.070           17.677          28.00
   h1       h2        S           b              d
   mm       mm        m           mm             mm
   75      180      3.003        250            700
    y       db       qd          qb              p
   mm       mm    kN/sq.m     kN/sq.m            m
  -41      266     2.00         7.20           17.280
    A       qD        ft          fb            Vol.
  sq.m   kN/sq.m   N/sq.mm     N/sq.mm          cu.m
  230     2.523    -1.865      -4.886           24.6
    s       Nt        Mf          Nf             Qf
    m      kN/m     kNm/m        kN/m           kN/m
 0.00    -520.8     5.76       -43.53          23.15
 0.45    -372.9    -1.96       -23.54          11.76
 0.90    -234.0    -5.39        -8.72           4.08
 1.35    -124.3    -6.11         0.93          -0.41
 1.80     -48.4    -5.37         6.22          -2.53
 2.25      -2.5    -4.06         8.26          -3.12
 2.70      20.4    -2.69         8.18          -2.86
 3.15      27.9    -1.54         6.91          -2.22
 3.60      26.7    -0.70         5.15          -1.51
 4.05      21.3    -0.17         3.36          -0.89
 4.95       9.1     0.25         0.61          -0.13
 6.30      -0.2     0.16        -1.01           0.13
```

TABLE 13 BP

DIA.	RISE	RADIUS	ALPHA
m	m	m	deg.
17.100	2.132	18.210	28.00

h1	h2	S	b	d
mm	mm	m	mm	mm
75	180	3.047	250	700

y	db	qd	qb	p
mm	mm	kN/sq.m	kN/sq.m	m
-41	266	1.99	6.79	17.800

A	qD	ft	fb	Vol.
sq.m	kN/sq.m	N/sq.mm	N/sq.mm	cu.m
244	2.513	-1.864	-4.860	26.0

s	Nt	Mf	Nf	Qf
m	kN/m	kNm/m	kN/m	kN/m
0.00	-519.2	5.57	-42.51	22.61
0.45	-373.0	-1.98	-23.11	11.57
0.90	-235.8	-5.38	-8.70	4.09
1.35	-126.9	-6.12	0.73	-0.32
1.80	-50.9	-5.43	5.97	-2.45
2.25	-4.4	-4.14	8.06	-3.08
2.70	19.4	-2.78	8.06	-2.86
3.15	27.8	-1.63	6.88	-2.25
3.60	27.1	-0.77	5.20	-1.56
4.05	22.1	-0.21	3.46	-0.94
5.40	5.2	0.26	-0.11	0.02
6.30	0.1	0.18	-0.93	0.13

```
TABLE  14  BP
  DIA.          RISE          RADIUS         ALPHA
   m             m              m            deg.
17.600         2.194         18.742         28.00
  h1           h2         S            b             d
  mm           mm         m            mm           mm
  75          180        3.092         250          700
  y           db        qd            qb            p
  mm          mm      kN/sq.m      kN/sq.m          m
 -41         266       1.99          6.41        18.321
  A           qD         ft           fb          Vol.
 sq.m      kN/sq.m    N/sq.mm      N/sq.mm        cu.m
 258        2.504     -1.864        -4.833         27.4
  s           Nt         Mf           Nf            Qf
  m          kN/m      kNm/m         kN/m          kN/m
0.00       -517.6      5.39        -41.54         22.09
0.45       -373.1     -2.01        -22.71         11.38
0.90       -237.5     -5.37         -8.69          4.10
1.35       -129.4     -6.14          0.55         -0.24
1.80        -53.3     -5.48          5.73         -2.37
2.25         -6.2     -4.22          7.85         -3.03
2.70         18.4     -2.88          7.94         -2.85
3.15         27.6     -1.71          6.85         -2.27
3.60         27.5     -0.84          5.24         -1.60
4.05         22.8     -0.26          3.55         -0.99
4.50         16.6      0.07          2.04         -0.52
5.40          5.9      0.27         -0.01          0.00
6.75         -0.8      0.13         -0.96          0.12
```

TABLE 15 BP

DIA. m		RISE m	RADIUS m	ALPHA deg.
18.000		2.244	19.168	28.00

h1 mm	h2 mm	S m	b mm	d mm
90	200	3.334	280	800

y mm	db mm	qd kN/sq.m	qb kN/sq.m	p m
-6	343	2.37	8.82	18.737

A sq.m	qD kN/sq.m	ft N/sq.mm	fb N/sq.mm	Vol. cu.m
270	2.929	-1.463	-4.067	33.6

s m	Nt kN/m	Mf kNm/m	Nf kN/m	Qf kN/m
0.00	-456.9	4.55	-37.34	19.86
0.45	-334.3	-2.18	-21.04	10.57
0.90	-219.4	-5.38	-8.67	4.10
1.35	-126.1	-6.24	-0.21	0.09
1.80	-58.1	-5.75	4.85	-2.02
2.25	-13.7	-4.62	7.24	-2.82
2.70	11.6	-3.33	7.76	-2.81
3.15	23.1	-2.15	7.09	-2.38
3.60	25.7	-1.21	5.79	-1.80
4.05	23.1	-0.53	4.28	-1.22
4.50	18.3	-0.09	2.81	-0.73
5.85	4.5	0.27	-0.16	0.03
6.75	0.2	0.18	-0.91	0.12

TABLE 16 BP

DIA.	RISE		RADIUS	ALPHA
m	m		m	deg.
18.500	2.307		19.701	28.00
h1	h2	S	b	d
mm	mm	m	mm	mm
90	200	3.380	280	800
y	db	qd	qb	p
mm	mm	kN/sq.m	kN/sq.m	m
-6	343	2.36	8.35	19.258
A	qD	ft	fb	Vol.
sq.m	kN/sq.m	N/sq.mm	N/sq.mm	cu.m
286	2.920	-1.464	-4.046	35.3
s	Nt	Mf	Nf	Qf
m	kN/m	kNm/m	kN/m	kN/m
0.00	-455.7	4.40	-36.55	19.44
0.45	-334.4	-2.21	-20.69	10.41
0.90	-220.8	-5.38	-8.63	4.10
1.35	-128.2	-6.25	-0.35	0.16
1.80	-60.3	-5.79	4.64	-1.95
2.25	-15.5	-4.69	7.05	-2.77
2.70	10.5	-3.42	7.62	-2.80
3.15	22.7	-2.24	7.02	-2.40
3.60	25.7	-1.28	5.79	-1.83
4.05	23.6	-0.59	4.33	-1.26
4.50	18.9	-0.14	2.90	-0.77
5.85	5.0	0.27	-0.06	0.01
7.20	-0.6	0.14	-0.96	0.12

TABLE 17 BP

DIA.	RISE	RADIUS	ALPHA
m	m	m	deg.
19.100	2.381	20.340	28.00

h1	h2	S	b	d
mm	mm	m	mm	mm
90	200	3.435	280	800

y	db	qd	qb	p
mm	mm	kN/sq.m	kN/sq.m	m
-6	343	2.36	7.83	19.882

A	qD	ft	fb	Vol.
sq.m	kN/sq.m	N/sq.mm	N/sq.mm	cu.m
304	2.909	-1.465	-4.021	37.5

s	Nt	Mf	Nf	Qf
m	kN/m	kNm/m	kN/m	kN/m
0.00	-454.3	4.22	-35.64	18.96
0.45	-334.6	-2.24	-20.30	10.22
0.90	-222.5	-5.37	-8.59	4.09
1.35	-130.6	-6.27	-0.51	0.23
1.80	-62.8	-5.84	4.41	-1.87
2.25	-17.5	-4.77	6.83	-2.71
2.70	9.2	-3.52	7.47	-2.78
3.15	22.1	-2.34	6.95	-2.41
3.60	25.8	-1.37	5.79	-1.87
4.05	24.1	-0.66	4.39	-1.31
4.50	19.7	-0.18	3.00	-0.82
4.95	14.5	0.10	1.77	-0.44
5.85	5.7	0.27	0.05	-0.01
7.20	-0.4	0.15	-0.89	0.12

TABLE 18 BP

DIA.	RISE		RADIUS	ALPHA
m	m		m	deg.
19.700	2.456		20.979	28.00

h1	h2	S	b	d
mm	mm	m	mm	mm
90	200	3.488	280	800

y	db	qd	qb	p
mm	mm	kN/sq.m	kN/sq.m	m
-6	343	2.36	7.36	20.507

A	qD	ft	fb	Vol.
sq.m	kN/sq.m	N/sq.mm	N/sq.mm	cu.m
324	2.898	-1.467	-3.996	39.8

s	Nt	Mf	Nf	Qf
m	kN/m	kNm/m	kN/m	kN/m
0.00	-453.0	4.05	-34.79	18.50
0.45	-334.7	-2.27	-19.92	10.05
0.90	-224.0	-5.37	-8.55	4.09
1.35	-132.9	-6.28	-0.66	0.30
1.80	-65.2	-5.89	4.19	-1.78
2.25	-19.6	-4.85	6.62	-2.66
2.70	7.8	-3.61	7.31	-2.75
3.15	21.4	-2.44	6.87	-2.42
3.60	25.8	-1.46	5.79	-1.90
4.05	24.5	-0.73	4.44	-1.35
4.50	20.4	-0.24	3.09	-0.87
4.95	15.3	0.06	1.88	-0.49
6.30	3.2	0.26	-0.35	0.07
7.65	-0.9	0.11	-0.86	0.11

TABLE 19 BP

DIA.	RISE	RADIUS	ALPHA
m	m	m	deg.
20.300	2.531	21.617	28.00

h1	h2	S	b	d
mm	mm	m	mm	mm
90	200	3.541	280	800

y	db	qd	qb	p
mm	mm	kN/sq.m	kN/sq.m	m
-6	343	2.35	6.93	21.131

A	qD	ft	fb	Vol.
sq.m	kN/sq.m	N/sq.mm	N/sq.mm	cu.m
344	2.888	-1.469	-3.972	42.1

s	Nt	Mf	Nf	Qf
m	kN/m	kNm/m	kN/m	kN/m
0.00	-451.6	3.90	-33.98	18.07
0.45	-334.8	-2.30	-19.56	9.89
0.90	-225.5	-5.36	-8.51	4.08
1.35	-135.2	-6.29	-0.80	0.37
1.80	-67.6	-5.93	3.97	-1.71
2.25	-21.6	-4.93	6.42	-2.60
2.70	6.4	-3.71	7.16	-2.73
3.15	20.7	-2.54	6.79	-2.43
3.60	25.7	-1.55	5.78	-1.93
4.05	24.9	-0.81	4.49	-1.40
4.50	21.0	-0.29	3.17	-0.91
4.95	16.1	0.03	1.99	-0.53
6.30	3.7	0.27	-0.26	0.05
7.65	-0.8	0.12	-0.81	0.11

TABLE 20 BP

DIA.		RISE		RADIUS		ALPHA
m		m		m		deg.
20.900		2.606		22.256		28.00

h1	h2	S	b	d
mm	mm	m	mm	mm
90	200	3.593	280	800

y	db	qd	qb	p
mm	mm	kN/sq.m	kN/sq.m	m
-6	343	2.35	6.54	21.756

A	qD	ft	fb	Vol.
sq.m	kN/sq.m	N/sq.mm	N/sq.mm	cu.m
364	2.878	-1.470	-3.948	44.4

s	Nt	Mf	Nf	Qf
m	kN/m	kNm/m	kN/m	kN/m
0.00	-450.2	3.75	-33.22	17.67
0.45	-334.8	-2.33	-19.22	9.73
0.90	-226.9	-5.35	-8.47	4.07
1.35	-137.4	-6.29	-0.94	0.43
1.80	-69.9	-5.97	3.77	-1.63
2.25	-23.6	-5.00	6.22	-2.54
2.70	5.0	-3.80	7.02	-2.70
3.15	20.0	-2.63	6.71	-2.43
3.60	25.6	-1.64	5.77	-1.96
4.05	25.2	-0.88	4.53	-1.44
4.50	21.7	-0.34	3.25	-0.96
4.95	16.8	0.00	2.08	-0.57
5.40	11.9	0.18	1.11	-0.28
6.75	1.9	0.24	-0.51	0.10
8.10	-1.0	0.09	-0.75	0.09

TABLE 21 BP

DIA.	RISE	RADIUS	ALPHA
m	m	m	deg.
21.500	2.681	22.895	28.00

h1	h2	S	b	d
mm	mm	m	mm	mm
90	200	3.644	280	800

y	db	qd	qb	p
mm	mm	kN/sq.m	kN/sq.m	m
-6	343	2.35	6.18	22.380

A	qD	ft	fb	Vol.
sq.m	kN/sq.m	N/sq.mm	N/sq.mm	cu.m
386	2.869	-1.472	-3.924	46.9

s	Nt	Mf	Nf	Qf
m	kN/m	kNm/m	kN/m	kN/m
0.00	-448.8	3.61	-32.50	17.28
0.45	-334.9	-2.35	-18.90	9.58
0.90	-228.2	-5.34	-8.43	4.07
1.35	-139.4	-6.30	-1.07	0.49
1.80	-72.2	-6.01	3.58	-1.55
2.25	-25.6	-5.07	6.03	-2.48
2.70	3.6	-3.88	6.88	-2.67
3.15	19.2	-2.72	6.63	-2.43
3.60	25.4	-1.73	5.75	-1.98
4.05	25.5	-0.95	4.57	-1.47
4.50	22.2	-0.40	3.32	-1.00
4.95	17.5	-0.04	2.17	-0.61
5.40	12.6	0.16	1.21	-0.31
6.75	2.3	0.25	-0.44	0.09
8.10	-1.0	0.10	-0.72	0.10

TABLE 22 BP

DIA.		RISE	RADIUS	ALPHA
m		m	m	deg.
22.100		2.755	23.534	28.00
h1	h2	S	b	d
mm	mm	m	mm	mm
90	200	3.695	280	800
y	db	qd	qb	p
mm	mm	kN/sq.m	kN/sq.m	m
-6	343	2.34	5.85	23.005
A	qD	ft	fb	Vol.
sq.m	kN/sq.m	N/sq.mm	N/sq.mm	cu.m
408	2.860	-1.474	-3.900	49.4
s	Nt	Mf	Nf	Qf
m	kN/m	kNm/m	kN/m	kN/m
0.00	-447.4	3.48	-31.81	16.91
0.45	-334.9	-2.37	-18.59	9.43
0.90	-229.5	-5.33	-8.40	4.06
1.35	-141.5	-6.30	-1.19	0.55
1.80	-74.4	-6.04	3.39	-1.48
2.25	-27.5	-5.13	5.85	-2.43
2.70	2.2	-3.97	6.74	-2.64
3.15	18.4	-2.81	6.55	-2.43
3.60	25.2	-1.81	5.73	-2.00
4.05	25.7	-1.02	4.60	-1.51
4.50	22.7	-0.45	3.39	-1.04
4.95	18.2	-0.08	2.26	-0.65
5.40	13.4	0.14	1.30	-0.34
6.75	2.7	0.25	-0.38	0.08
8.55	-1.1	0.07	-0.65	0.08

TABLE 23 BP

DIA.	RISE	RADIUS	ALPHA
m	m	m	deg.
22.800	2.843	24.280	28.00

h1	h2	S	b	d
mm	mm	m	mm	mm
90	200	3.753	280	800

y	db	qd	qb	p
mm	mm	kN/sq.m	kN/sq.m	m
-6	343	2.34	5.50	23.734

A	qD	ft	fb	Vol.
sq.m	kN/sq.m	N/sq.mm	N/sq.mm	cu.m
434	2.850	-1.477	-3.873	52.4

s	Nt	Mf	Nf	Qf
m	kN/m	kNm/m	kN/m	kN/m
0.00	-445.8	3.33	-31.04	16.51
0.45	-334.8	-2.39	-18.25	9.27
0.90	-230.9	-5.31	-8.35	4.05
1.35	-143.7	-6.30	-1.32	0.61
1.80	-76.9	-6.08	3.18	-1.40
2.25	-29.7	-5.20	5.65	-2.36
2.70	0.5	-4.06	6.58	-2.61
3.15	17.4	-2.92	6.46	-2.42
3.60	24.8	-1.91	5.70	-2.02
4.05	25.9	-1.11	4.63	-1.54
4.50	23.3	-0.52	3.46	-1.08
4.95	19.0	-0.12	2.36	-0.69
5.40	14.2	0.12	1.40	-0.38
5.85	9.7	0.23	0.64	-0.16
7.20	1.2	0.22	-0.53	0.10
8.55	-1.1	0.08	-0.63	0.09

TABLE 24 BP

DIA. m	RISE m	RADIUS m	ALPHA deg.
23.500	2.930	25.025	28.00

h1 mm	h2 mm	S m	b mm	d mm
90	200	3.810	280	800

y mm	db mm	qd kN/sq.m	qb kN/sq.m	p m
-6	343	2.34	5.17	24.462

A sq.m	qD kN/sq.m	ft N/sq.mm	fb N/sq.mm	Vol. cu.m
461	2.840	-1.479	-3.846	55.4

s m	Nt kN/m	Mf kNm/m	Nf kN/m	Qf kN/m
0.00	-444.3	3.20	-30.32	16.13
0.45	-334.8	-2.41	-17.92	9.12
0.90	-232.2	-5.30	-8.31	4.04
1.35	-145.9	-6.30	-1.45	0.67
1.80	-79.3	-6.11	2.99	-1.32
2.25	-31.9	-5.26	5.45	-2.30
2.70	-1.2	-4.15	6.42	-2.57
3.15	16.4	-3.01	6.37	-2.42
3.60	24.4	-2.01	5.67	-2.04
4.05	26.0	-1.19	4.65	-1.58
4.50	23.8	-0.59	3.52	-1.12
4.95	19.7	-0.17	2.44	-0.73
5.40	14.9	0.09	1.50	-0.42
5.85	10.5	0.22	0.74	-0.19
7.20	1.6	0.23	-0.47	0.10
9.00	-1.1	0.05	-0.54	0.07

TABLE 25 BP

DIA.		RISE		RADIUS		ALPHA
m		m		m		deg.
24.000		2.992		25.558		28.00

h1	h2	S	b	d
mm	mm	m	mm	mm
100	250	4.230	300	1000

y	db	qd	qb	p
mm	mm	kN/sq.m	kN/sq.m	m
43	487	2.68	6.12	24.983

A	qD	ft	fb	Vol.
sq.m	kN/sq.m	N/sq.mm	N/sq.mm	cu.m
481	3.411	-1.039	-3.018	69.5

s	Nt	Mf	Nf	Qf
m	kN/m	kNm/m	kN/m	kN/m
0.00	-395.5	2.70	-28.80	15.32
0.45	-306.0	-2.74	-17.95	9.14
0.90	-221.7	-5.75	-9.25	4.51
1.35	-148.9	-7.01	-2.72	1.27
1.80	-90.2	-7.07	1.84	-0.82
2.25	-45.8	-6.41	4.71	-1.99
2.70	-14.5	-5.37	6.22	-2.51
3.15	5.8	-4.22	6.71	-2.57
3.60	17.5	-3.10	6.48	-2.35
4.05	22.9	-2.12	5.80	-1.99
4.50	23.8	-1.32	4.86	-1.57
4.95	22.0	-0.71	3.83	-1.16
5.40	18.8	-0.27	2.82	-0.80
5.85	15.0	0.03	1.91	-0.51
7.65	2.8	0.29	-0.35	0.07
9.00	-0.6	0.16	-0.78	0.11

```
TABLE 26 BP
 DIA.          RISE           RADIUS         ALPHA
   m             m              m            deg.
24.700         3.080         26.303         28.00
  h1       h2        S            b            d
  mm       mm        m           mm           mm
 100      250        4.291       300         1000
   y       db        qd           qb           p
  mm       mm     kN/sq.m      kN/sq.m         m
  43      487       2.67         5.78        25.712
   A       qD         ft           fb         Vol.
 sq.m   kN/sq.m    N/sq.mm      N/sq.mm       cu.m
 509     3.398     -1.042       -2.997        73.3
   s       Nt         Mf           Nf           Qf
   m      kN/m      kNm/m         kN/m         kN/m
0.00    -394.3      2.57        -28.18        14.98
0.45    -306.0     -2.77        -17.63         8.99
0.90    -222.7     -5.74         -9.18         4.48
1.35    -150.6     -7.00         -2.80         1.31
1.80     -92.2     -7.09          1.67        -0.75
2.25     -47.8     -6.46          4.51        -1.93
2.70     -16.2     -5.45          6.04        -2.46
3.15       4.6     -4.32          6.57        -2.54
3.60      16.8     -3.21          6.40        -2.35
4.05      22.6     -2.22          5.76        -2.01
4.50      23.9     -1.41          4.87        -1.60
4.95      22.4     -0.78          3.88        -1.20
5.40      19.4     -0.32          2.90        -0.84
5.85      15.6     -0.01          1.99        -0.54
6.30      11.9      0.18          1.22        -0.31
7.65       3.2      0.30         -0.27         0.06
9.45      -0.8      0.12         -0.74         0.10
```

TABLE 27 BP

DIA.	RISE	RADIUS	ALPHA
m	m	m	deg.
25.500	3.179	27.155	28.00

h1	h2	S	b	d
mm	mm	m	mm	mm
100	250	4.360	300	1000

y	db	qd	qb	p
mm	mm	kN/sq.m	kN/sq.m	m
43	487	2.67	5.42	26.544

A	qD	ft	fb	Vol.
sq.m	kN/sq.m	N/sq.mm	N/sq.mm	cu.m
543	3.383	-1.045	-2.975	77.8

s	Nt	Mf	Nf	Qf
m	kN/m	kNm/m	kN/m	kN/m
0.00	-393.0	2.43	-27.50	14.62
0.45	-305.9	-2.79	-17.30	8.83
0.90	-223.8	-5.73	-9.10	4.46
1.35	-152.5	-7.00	-2.89	1.36
1.80	-94.5	-7.11	1.49	-0.67
2.25	-50.1	-6.52	4.31	-1.85
2.70	-18.2	-5.54	5.85	-2.40
3.15	3.1	-4.42	6.42	-2.51
3.60	15.8	-3.32	6.30	-2.34
4.05	22.2	-2.34	5.72	-2.02
4.50	24.0	-1.51	4.88	-1.63
4.95	22.8	-0.87	3.93	-1.24
5.40	20.0	-0.39	2.97	-0.89
5.85	16.4	-0.06	2.09	-0.59
6.30	12.6	0.15	1.32	-0.35
8.10	1.9	0.28	-0.44	0.09
9.90	-1.0	0.09	-0.68	0.08

TABLE 28 BP

DIA.	RISE	RADIUS	ALPHA
m	m	m	deg.
26.200	3.267	27.900	28.00

h1	h2	S	b	d
mm	mm	m	mm	mm
100	250	4.419	300	1000

y	db	qd	qb	p
mm	mm	kN/sq.m	kN/sq.m	m
43	487	2.66	5.14	27.273

A	qD	ft	fb	Vol.
sq.m	kN/sq.m	N/sq.mm	N/sq.mm	cu.m
573	3.371	-1.047	-2.955	81.8

s	Nt	Mf	Nf	Qf
m	kN/m	kNm/m	kN/m	kN/m
0.00	-391.8	2.31	-26.94	14.33
0.45	-305.8	-2.81	-17.02	8.70
0.90	-224.7	-5.72	-9.03	4.43
1.35	-154.1	-6.99	-2.96	1.40
1.80	-96.4	-7.13	1.34	-0.61
2.25	-52.0	-6.57	4.13	-1.79
2.70	-19.9	-5.61	5.69	-2.35
3.15	1.8	-4.51	6.29	-2.48
3.60	15.0	-3.42	6.22	-2.34
4.05	21.8	-2.43	5.68	-2.03
4.50	24.0	-1.60	4.89	-1.66
4.95	23.1	-0.94	3.97	-1.28
5.40	20.5	-0.45	3.04	-0.92
5.85	17.0	-0.10	2.17	-0.62
6.30	13.2	0.12	1.40	-0.38
6.75	9.7	0.25	0.77	-0.19
8.10	2.3	0.28	-0.38	0.08
9.90	-0.9	0.11	-0.65	0.09

TABLE 29 BP

DIA.	RISE	RADIUS	ALPHA
m	m	m	deg.
27.000	3.366	28.752	28.00

h1	h2	S	b	d
mm	mm	m	mm	mm
100	250	4.486	300	1000

y	db	qd	qb	p
mm	mm	kN/sq.m	kN/sq.m	m
43	487	2.66	4.84	28.106

A	qD	ft	fb	Vol.
sq.m	kN/sq.m	N/sq.mm	N/sq.mm	cu.m
608	3.358	-1.050	-2.933	86.5

s	Nt	Mf	Nf	Qf
m	kN/m	kNm/m	kN/m	kN/m
0.00	-390.4	2.19	-26.33	14.00
0.45	-305.6	-2.84	-16.71	8.56
0.90	-225.7	-5.70	-8.95	4.41
1.35	-155.9	-6.98	-3.04	1.44
1.80	-98.6	-7.15	1.18	-0.54
2.25	-54.2	-6.61	3.94	-1.71
2.70	-21.8	-5.69	5.51	-2.29
3.15	0.3	-4.61	6.15	-2.45
3.60	14.0	-3.53	6.12	-2.33
4.05	21.3	-2.54	5.64	-2.04
4.50	23.9	-1.70	4.88	-1.69
4.95	23.4	-1.03	4.01	-1.31
5.40	21.0	-0.52	3.10	-0.96
5.85	17.6	-0.15	2.25	-0.66
6.30	14.0	0.09	1.50	-0.41
6.75	10.4	0.23	0.86	-0.22
8.55	1.2	0.26	-0.49	0.10
10.35	-1.0	0.08	-0.59	0.08

TABLE 30 BP

DIA.		RISE		RADIUS		ALPHA
m		m		m		deg.
27.800		3.466		29.604		28.00

h1	h2	S	b	d
mm	mm	m	mm	mm
100	250	4.552	300	1000

y	db	qd	qb	p
mm	mm	kN/sq.m	kN/sq.m	m
43	487	2.66	4.56	28.938

A	qD	ft	fb	Vol.
sq.m	kN/sq.m	N/sq.mm	N/sq.mm	cu.m
645	3.345	-1.053	-2.911	91.4

s	Nt	Mf	Nf	Qf
m	kN/m	kNm/m	kN/m	kN/m
0.00	-389.1	2.07	-25.75	13.69
0.45	-305.5	-2.85	-16.42	8.42
0.90	-226.6	-5.69	-8.88	4.38
1.35	-157.6	-6.97	-3.11	1.48
1.80	-100.6	-7.16	1.03	-0.47
2.25	-56.3	-6.66	3.76	-1.65
2.70	-23.6	-5.77	5.33	-2.24
3.15	-1.2	-4.71	6.01	-2.42
3.60	13.0	-3.63	6.02	-2.32
4.05	20.8	-2.65	5.59	-2.05
4.50	23.8	-1.80	4.88	-1.71
4.95	23.6	-1.11	4.04	-1.35
5.40	21.5	-0.59	3.16	-1.00
5.85	18.2	-0.21	2.33	-0.70
6.30	14.6	0.05	1.58	-0.45
6.75	11.1	0.21	0.95	-0.25
7.20	7.9	0.29	0.43	-0.11
9.00	0.4	0.22	-0.55	0.10
10.80	-1.0	0.06	-0.52	0.06

TABLE 31 BP

DIA.	RISE	RADIUS	ALPHA
m	m	m	deg.
28.700	3.578	30.563	28.00

h1	h2	S	b	d
mm	mm	m	mm	mm
100	250	4.625	300	1000

y	db	qd	qb	p
mm	mm	kN/sq.m	kN/sq.m	m
43	487	2.65	4.28	29.875

A	qD	ft	fb	Vol.
sq.m	kN/sq.m	N/sq.mm	N/sq.mm	cu.m
687	3.331	-1.056	-2.887	97.0

s	Nt	Mf	Nf	Qf
m	kN/m	kNm/m	kN/m	kN/m
0.00	-387.6	1.95	-25.14	13.37
0.45	-305.3	-2.87	-16.12	8.27
0.90	-227.6	-5.67	-8.81	4.35
1.35	-159.4	-6.95	-3.19	1.52
1.80	-102.9	-7.17	0.86	-0.40
2.25	-58.6	-6.71	3.56	-1.57
2.70	-25.7	-5.84	5.14	-2.18
3.15	-2.8	-4.81	5.85	-2.38
3.60	11.9	-3.75	5.91	-2.30
4.05	20.1	-2.76	5.53	-2.06
4.50	23.6	-1.91	4.87	-1.73
4.95	23.8	-1.21	4.07	-1.38
5.40	21.9	-0.66	3.22	-1.04
5.85	18.9	-0.27	2.41	-0.74
6.30	15.4	0.01	1.68	-0.49
6.75	11.9	0.18	1.04	-0.29
7.20	8.6	0.27	0.52	-0.14
9.00	0.7	0.24	-0.50	0.10
10.80	-1.0	0.07	-0.51	0.07

```
TABLE 32 BP
 DIA.         RISE         RADIUS        ALPHA
  m            m             m            deg.
29.500       3.678        31.415        28.00
  h1          h2          S             b            d
  mm          mm          m             mm           mm
 100         250           4.689        300         1000
  y           db          qd            qb           p
  mm          mm       kN/sq.m       kN/sq.m         m
  43         487          2.65          4.05        30.708
  A           qD          ft            fb          Vol.
 sq.m      kN/sq.m     N/sq.mm       N/sq.mm        cu.m
 726         3.319       -1.059        -2.866       102.1
  s           Nt          Mf            Nf           Qf
  m          kN/m        kNm/m         kN/m         kN/m
 0.00       -386.2        1.85        -24.62        13.09
 0.45       -305.0       -2.88        -15.85         8.14
 0.90       -228.4       -5.65         -8.74         4.33
 1.35       -160.9       -6.93         -3.26         1.56
 1.80       -104.8       -7.18          0.72        -0.33
 2.25        -60.6       -6.74          3.39        -1.50
 2.70        -27.5       -5.91          4.98        -2.12
 3.15         -4.3       -4.89          5.72        -2.34
 3.60         10.8       -3.84          5.82        -2.28
 4.05         19.5       -2.86          5.48        -2.06
 4.50         23.4       -2.00          4.86        -1.75
 4.95         23.9       -1.29          4.09        -1.41
 5.40         22.3       -0.74          3.27        -1.07
 5.85         19.4       -0.32          2.48        -0.77
 6.30         16.0       -0.03          1.75        -0.52
 6.75         12.5        0.15          1.12        -0.32
 7.20          9.3        0.26          0.60        -0.16
 7.65          6.4        0.30          0.19        -0.05
 9.45          0.0        0.20         -0.53         0.10
11.25         -1.0        0.05         -0.44         0.06
```

TABLE 33 BP

DIA.	RISE	RADIUS	ALPHA
m	m	m	deg.
30.000	3.740	31.947	28.00

h1	h2	S	b	d
mm	mm	m	mm	mm
130	290	5.180	300	1200

y	db	qd	qb	p
mm	mm	kN/sq.m	kN/sq.m	m
115	642	3.40	6.62	31.229

A	qD	ft	fb	Vol.
sq.m	kN/sq.m	N/sq.mm	N/sq.mm	cu.m
751	4.169	-0.832	-2.483	132.7

s	Nt	Mf	Nf	Qf
m	kN/m	kNm/m	kN/m	kN/m
0.00	-361.1	0.87	-23.74	12.63
0.45	-289.0	-3.75	-15.66	8.05
0.90	-221.0	-6.53	-9.00	4.46
1.35	-160.7	-7.90	-3.73	1.79
1.80	-109.6	-8.25	0.23	-0.11
2.25	-68.2	-7.90	3.03	-1.35
2.70	-36.1	-7.12	4.84	-2.07
3.15	-12.4	-6.10	5.84	-2.40
3.60	4.2	-5.00	6.21	-2.45
4.05	14.8	-3.92	6.11	-2.31
4.50	20.7	-2.93	5.67	-2.06
4.95	23.2	-2.08	5.03	-1.74
5.40	23.2	-1.36	4.28	-1.42
5.85	21.6	-0.80	3.48	-1.10
6.30	19.0	-0.37	2.71	-0.81
6.75	15.9	-0.06	1.99	-0.57
7.20	12.8	0.15	1.35	-0.37
7.65	9.8	0.27	0.81	-0.21
9.45	1.7	0.31	-0.45	0.09
11.70	-1.0	0.09	-0.62	0.08

TABLE 34 BP

DIA. m	RISE m		RADIUS m	ALPHA deg.
30.900	3.853		32.905	28.00

h1 mm	h2 mm	S m	b mm	d mm
130	290	5.257	300	1200

y mm	db mm	qd kN/sq.m	qb kN/sq.m	p m
115	642	3.39	6.24	32.165

A sq.m	qD kN/sq.m	ft N/sq.mm	fb N/sq.mm	Vol. cu.m
797	4.155	-0.835	-2.463	140.3

s m	Nt kN/m	Mf kNm/m	Nf kN/m	Qf kN/m
0.00	-359.9	0.76	-23.23	12.35
0.45	-288.8	-3.76	-15.40	7.92
0.90	-221.8	-6.51	-8.92	4.43
1.35	-162.2	-7.88	-3.79	1.82
1.80	-111.5	-8.25	0.09	-0.04
2.25	-70.2	-7.94	2.85	-1.27
2.70	-38.0	-7.18	4.65	-2.00
3.15	-14.1	-6.19	5.67	-2.35
3.60	2.8	-5.11	6.07	-2.42
4.05	13.8	-4.04	6.01	-2.30
4.50	20.2	-3.05	5.62	-2.06
4.95	23.0	-2.19	5.02	-1.77
5.40	23.3	-1.47	4.30	-1.45
5.85	21.9	-0.89	3.53	-1.14
6.30	19.5	-0.44	2.78	-0.85
6.75	16.6	-0.12	2.07	-0.60
7.20	13.5	0.11	1.44	-0.40
7.65	10.5	0.25	0.90	-0.24
9.90	0.9	0.28	-0.52	0.10
11.70	-1.0	0.11	-0.60	0.08

TABLE 35 BP

DIA.	RISE	RADIUS	ALPHA
m	m	m	deg.
31.800	3.965	33.864	28.00

h1	h2	S	b	d
mm	mm	m	mm	mm
130	290	5.333	300	1200

y	db	qd	qb	p
mm	mm	kN/sq.m	kN/sq.m	m
115	642	3.39	5.89	33.102

A	qD	ft	fb	Vol.
sq.m	kN/sq.m	N/sq.mm	N/sq.mm	cu.m
844	4.141	-0.838	-2.443	148.0

s	Nt	Mf	Nf	Qf
m	kN/m	kNm/m	kN/m	kN/m
0.00	-358.7	0.67	-22.74	12.09
0.45	-288.6	-3.77	-15.14	7.79
0.90	-222.6	-6.48	-8.84	4.41
1.35	-163.6	-7.86	-3.84	1.85
1.80	-113.3	-8.25	-0.04	0.02
2.25	-72.2	-7.97	2.68	-1.20
2.70	-40.0	-7.24	4.48	-1.94
3.15	-15.8	-6.28	5.51	-2.30
3.60	1.5	-5.21	5.94	-2.39
4.05	12.8	-4.15	5.91	-2.29
4.50	19.6	-3.17	5.56	-2.07
4.95	22.8	-2.30	5.00	-1.78
5.40	23.4	-1.57	4.31	-1.47
5.85	22.2	-0.98	3.58	-1.17
6.30	20.0	-0.51	2.84	-0.89
6.75	17.2	-0.17	2.15	-0.64
7.20	14.1	0.07	1.53	-0.43
7.65	11.1	0.22	0.99	-0.27
8.10	8.4	0.31	0.53	-0.14
9.90	1.2	0.29	-0.47	0.09
12.15	-1.0	0.08	-0.54	0.07

TABLE 36 BP

DIA.		RISE		RADIUS		ALPHA
m		m		m		deg.
32.800		4.090		34.929		28.00
h1	h2		S		b	d
mm	mm		m		mm	mm
130	290		5.417		300	1200
y	db		qd		qb	p
mm	mm		kN/sq.m		kN/sq.m	m
115	642		3.39		5.54	34.143
A	qD		ft		fb	Vol.
sq.m	kN/sq.m		N/sq.mm		N/sq.mm	cu.m
898	4.126		-0.841		-2.421	157.0

s	Nt	Mf	Nf	Qf
m	kN/m	kNm/m	kN/m	kN/m
0.00	-357.4	0.57	-22.23	11.82
0.45	-288.4	-3.78	-14.87	7.66
0.90	-223.3	-6.46	-8.76	4.37
1.35	-165.1	-7.83	-3.89	1.88
1.80	-115.3	-8.25	-0.18	0.08
2.25	-74.4	-8.00	2.50	-1.13
2.70	-42.1	-7.31	4.29	-1.87
3.15	-17.6	-6.37	5.33	-2.25
3.60	-0.1	-5.32	5.80	-2.35
4.05	11.7	-4.28	5.81	-2.27
4.50	18.9	-3.30	5.50	-2.07
4.95	22.5	-2.42	4.98	-1.80
5.40	23.4	-1.68	4.33	-1.50
5.85	22.6	-1.07	3.62	-1.20
6.30	20.5	-0.59	2.91	-0.93
6.75	17.8	-0.23	2.23	-0.68
7.20	14.8	0.02	1.62	-0.47
7.65	11.8	0.19	1.08	-0.30
8.10	9.1	0.30	0.62	-0.16
10.35	0.6	0.27	-0.51	0.10
12.60	-1.0	0.07	-0.48	0.06

TABLE 37 BP

DIA.	RISE	RADIUS	ALPHA
m	m	m	deg.
33.800	4.214	35.994	28.00

h1	h2	S	b	d
mm	mm	m	mm	mm
130	290	5.499	300	1200

y	db	qd	qb	p
mm	mm	kN/sq.m	kN/sq.m	m
115	642	3.38	5.22	35.184

A	qD	ft	fb	Vol.
sq.m	kN/sq.m	N/sq.mm	N/sq.mm	cu.m
953	4.112	-0.843	-2.400	166.1

s	Nt	Mf	Nf	Qf
m	kN/m	kNm/m	kN/m	kN/m
0.00	-356.0	0.48	-21.74	11.56
0.45	-288.1	-3.79	-14.61	7.54
0.90	-224.0	-6.43	-8.68	4.34
1.35	-166.5	-7.81	-3.94	1.91
1.80	-117.2	-8.25	-0.31	0.14
2.25	-76.5	-8.02	2.33	-1.06
2.70	-44.1	-7.36	4.10	-1.80
3.15	-19.5	-6.45	5.16	-2.19
3.60	-1.6	-5.43	5.65	-2.32
4.05	10.6	-4.40	5.71	-2.25
4.50	18.2	-3.42	5.44	-2.07
4.95	22.1	-2.54	4.95	-1.81
5.40	23.4	-1.79	4.33	-1.53
5.85	22.8	-1.17	3.66	-1.24
6.30	20.9	-0.68	2.97	-0.96
6.75	18.4	-0.30	2.31	-0.72
7.20	15.5	-0.03	1.70	-0.50
7.65	12.5	0.16	1.16	-0.33
8.10	9.8	0.28	0.71	-0.19
8.55	7.3	0.34	0.33	-0.08
10.80	0.1	0.24	-0.52	0.10
13.05	-1.0	0.05	-0.42	0.05

TABLE 38 BP

DIA.	RISE	RADIUS	ALPHA
m	m	m	deg.
34.800	4.339	37.059	28.00

h1	h2	S	b	d
mm	mm	m	mm	mm
130	290	5.579	300	1200

y	db	qd	qb	p
mm	mm	kN/sq.m	kN/sq.m	m
115	642	3.38	4.92	36.225

A	qD	ft	fb	Vol.
sq.m	kN/sq.m	N/sq.mm	N/sq.mm	cu.m
1010	4.099	-0.846	-2.379	175.5

s	Nt	Mf	Nf	Qf
m	kN/m	kNm/m	kN/m	kN/m
0.00	-354.7	0.40	-21.28	11.32
0.45	-287.8	-3.79	-14.37	7.42
0.90	-224.6	-6.40	-8.61	4.31
1.35	-167.9	-7.78	-3.99	1.94
1.80	-119.0	-8.24	-0.43	0.20
2.25	-78.5	-8.04	2.16	-0.99
2.70	-46.1	-7.42	3.93	-1.74
3.15	-21.3	-6.53	5.00	-2.14
3.60	-3.1	-5.53	5.52	-2.28
4.05	9.5	-4.51	5.60	-2.23
4.50	17.4	-3.54	5.37	-2.07
4.95	21.7	-2.66	4.92	-1.83
5.40	23.3	-1.90	4.34	-1.55
5.85	23.0	-1.27	3.69	-1.27
6.30	21.3	-0.76	3.02	-1.00
6.75	18.9	-0.37	2.37	-0.75
7.20	16.1	-0.08	1.78	-0.54
7.65	13.2	0.12	1.24	-0.36
8.10	10.4	0.25	0.79	-0.22
8.55	7.9	0.32	0.41	-0.11
9.00	5.7	0.35	0.10	-0.03
11.25	-0.3	0.21	-0.52	0.10
13.50	-0.9	0.04	-0.37	0.05

TABLE 39 BP

DIA.	RISE	RADIUS	ALPHA
m	m	m	deg.
35.000	4.364	37.272	28.00

h1	h2	S	b	d
mm	mm	m	mm	mm
130	290	5.595	300	1200

y	db	qd	qb	p
mm	mm	kN/sq.m	kN/sq.m	m
115	642	3.38	4.87	36.433

A	qD	ft	fb	Vol.
sq.m	kN/sq.m	N/sq.mm	N/sq.mm	cu.m
1022	4.096	-0.847	-2.375	177.4

s	Nt	Mf	Nf	Qf
m	kN/m	kNm/m	kN/m	kN/m
0.00	-354.4	0.38	-21.19	11.27
0.45	-287.7	-3.79	-14.32	7.40
0.90	-224.8	-6.39	-8.59	4.31
1.35	-168.2	-7.77	-3.99	1.94
1.80	-119.4	-8.24	-0.46	0.21
2.25	-78.9	-8.05	2.13	-0.97
2.70	-46.5	-7.42	3.89	-1.72
3.15	-21.6	-6.55	4.97	-2.13
3.60	-3.4	-5.55	5.49	-2.27
4.05	9.2	-4.53	5.58	-2.23
4.50	17.3	-3.56	5.36	-2.07
4.95	21.7	-2.68	4.92	-1.83
5.40	23.3	-1.92	4.34	-1.55
5.85	23.0	-1.29	3.69	-1.27
6.30	21.4	-0.78	3.03	-1.00
6.75	19.0	-0.38	2.39	-0.76
7.20	16.2	-0.09	1.79	-0.55
7.65	13.3	0.12	1.26	-0.37
8.10	10.5	0.25	0.80	-0.22
8.55	8.0	0.32	0.42	-0.11
9.00	5.8	0.35	0.12	-0.03
11.25	-0.3	0.21	-0.52	0.10
13.50	-1.0	0.04	-0.37	0.05

```
TABLE 40 BP
```

DIA.	RISE	RADIUS	ALPHA
m	m	m	deg.
36.100	4.501	38.443	28.00

h1	h2	S	b	d
mm	mm	m	mm	mm
130	290	5.683	300	1200

y	db	qd	qb	p
mm	mm	kN/sq.m	kN/sq.m	m
115	642	3.37	4.57	37.578

A	qD	ft	fb	Vol.
sq.m	kN/sq.m	N/sq.mm	N/sq.mm	cu.m
1087	4.082	-0.850	-2.353	188.1

s	Nt	Mf	Nf	Qf
m	kN/m	kNm/m	kN/m	kN/m
0.00	-353.0	0.30	-20.72	11.02
0.45	-287.4	-3.79	-14.07	7.27
0.90	-225.4	-6.36	-8.51	4.27
1.35	-169.6	-7.74	-4.04	1.97
1.80	-121.3	-8.22	-0.58	0.28
2.25	-81.0	-8.06	1.96	-0.90
2.70	-48.6	-7.48	3.71	-1.65
3.15	-23.6	-6.63	4.80	-2.07
3.60	-5.0	-5.65	5.34	-2.23
4.05	8.0	-4.65	5.47	-2.21
4.50	16.4	-3.69	5.28	-2.06
4.95	21.2	-2.81	4.88	-1.84
5.40	23.2	-2.04	4.34	-1.57
5.85	23.2	-1.39	3.72	-1.30
6.30	21.8	-0.87	3.08	-1.04
6.75	19.6	-0.45	2.45	-0.79
7.20	16.9	-0.15	1.87	-0.58
7.65	14.0	0.07	1.35	-0.40
8.10	11.2	0.22	0.89	-0.25
8.55	8.7	0.31	0.50	-0.14
9.00	6.4	0.35	0.19	-0.05
11.70	-0.5	0.18	-0.50	0.09
13.95	-0.9	0.03	-0.32	0.04

Tables 1CP to 40CP   For Type C Domes

TABLE 1 CP

DIA.	RISE	RADIUS	ALPHA
m	m	m	deg.
12.000	1.496	12.779	28.00

h1	h2	S	b	d
mm	mm	m	mm	mm
75	180	2.553	250	700

y	db	qd	qb	p
mm	mm	kN/sq.m	kN/sq.m	m
-308	0	2.03	13.78	12.491

A	qD	ft	fb	Vol.
sq.m	kN/sq.m	N/sq.mm	N/sq.mm	cu.m
120	2.636	-3.600	-2.583	13.4

s	Nt	Mf	Nf	Qf
m	kN/m	kNm/m	kN/m	kN/m
0.00	-371.5	21.79	-67.53	35.91
0.45	-352.7	8.60	-46.83	22.83
0.90	-262.9	0.91	-26.74	11.89
1.35	-163.3	-2.62	-10.94	4.41
1.80	-82.5	-3.54	-0.44	0.16
2.25	-28.4	-3.12	5.31	-1.72
2.70	1.5	-2.22	7.47	-2.13
3.60	16.3	-0.62	5.99	-1.26
4.50	9.1	0.06	2.40	-0.33

TABLE 2 CP

DIA.	RISE	RADIUS	ALPHA
m	m	m	deg.
12.400	1.546	13.205	28.00
h1	h2	S	b	d
mm	mm	m	mm	mm
75	180	2.595	250	700
y	db	qd	qb	p
mm	mm	kN/sq.m	kN/sq.m	m
-308	0	2.03	12.90	12.908
A	qD	ft	fb	Vol.
sq.m	kN/sq.m	N/sq.mm	N/sq.mm	cu.m
128	2.624	-3.522	-2.598	14.3
s	Nt	Mf	Nf	Qf
m	kN/m	kNm/m	kN/m	kN/m
0.00	-374.0	21.52	-66.15	35.18
0.45	-353.6	8.58	-45.96	22.47
0.90	-265.0	0.97	-26.46	11.83
1.35	-166.4	-2.58	-11.08	4.51
1.80	-85.7	-3.55	-0.77	0.28
2.25	-31.0	-3.18	4.95	-1.63
2.70	-0.1	-2.30	7.20	-2.10
3.15	13.5	-1.39	7.18	-1.84
3.60	16.4	-0.68	5.94	-1.30
4.50	9.8	0.03	2.49	-0.37

TABLE 3 CP

DIA.		RISE		RADIUS	ALPHA
m		m		m	deg.
12.700		1.583		13.524	28.00
h1	h2		S	b	d
mm	mm		m	mm	mm
75	180		2.626	250	700
y	db	qd		qb	p
mm	mm	kN/sq.m		kN/sq.m	m
-308	0	2.02		12.30	13.220
A	qD	ft		fb	Vol.
sq.m	kN/sq.m	N/sq.mm		N/sq.mm	cu.m
135	2.615	-3.466		-2.608	14.9
s	Nt	Mf		Nf	Qf
m	kN/m	kNm/m		kN/m	kN/m
0.00	-375.7	21.31		-65.15	34.64
0.45	-354.3	8.55		-45.33	22.20
0.90	-266.4	1.01		-26.24	11.79
1.35	-168.6	-2.55		-11.16	4.57
1.80	-88.1	-3.56		-0.99	0.37
2.25	-32.9	-3.22		4.71	-1.57
2.70	-1.3	-2.35		7.00	-2.08
3.15	13.0	-1.45		7.07	-1.85
4.05	14.4	-0.25		4.24	-0.81
4.95	6.2	0.13		1.07	-0.13

TABLE 4 CP

DIA.	RISE		RADIUS	ALPHA
m	m		m	deg.
13.100	1.633		13.950	28.00
h1	h2	S	b	d
mm	mm	m	mm	mm
75	180	2.667	250	700
y	db	qd	qb	p
mm	mm	kN/sq.m	kN/sq.m	m
-308	0	2.02	11.56	13.636
A	qD	ft	fb	Vol.
sq.m	kN/sq.m	N/sq.mm	N/sq.mm	cu.m
143	2.604	-3.394	-2.621	15.8
s	Nt	Mf	Nf	Qf
m	kN/m	kNm/m	kN/m	kN/m
0.00	-377.9	21.04	-63.85	33.95
0.45	-355.1	8.51	-44.51	21.86
0.90	-268.2	1.06	-25.95	11.72
1.35	-171.4	-2.51	-11.25	4.65
1.80	-91.0	-3.56	-1.27	0.48
2.25	-35.4	-3.27	4.39	-1.49
2.70	-2.8	-2.43	6.75	-2.05
3.15	12.4	-1.53	6.92	-1.86
4.05	14.8	-0.30	4.27	-0.86
4.95	6.7	0.12	1.19	-0.16

```
TABLE 5 CP
```

DIA.	RISE	RADIUS	ALPHA
m	m	m	deg.
13.500	1.683	14.376	28.00

h1	h2	S	b	d
mm	mm	m	mm	mm
75	180	2.708	250	700

y	db	qd	qb	p
mm	mm	kN/sq.m	kN/sq.m	m
-308	0	2.02	10.89	14.053

A	qD	ft	fb	Vol.
sq.m	kN/sq.m	N/sq.mm	N/sq.mm	cu.m
152	2.594	-3.325	-2.634	16.7

s	Nt	Mf	Nf	Qf
m	kN/m	kNm/m	kN/m	kN/m
0.00	-380.1	20.76	-62.60	33.29
0.45	-355.8	8.46	-43.71	21.52
0.90	-269.9	1.10	-25.65	11.65
1.35	-174.0	-2.47	-11.32	4.72
1.80	-93.9	-3.57	-1.53	0.58
2.25	-37.8	-3.31	4.10	-1.42
2.70	-4.4	-2.50	6.52	-2.02
3.15	11.7	-1.60	6.77	-1.87
4.05	15.2	-0.34	4.30	-0.90
4.95	7.3	0.11	1.31	-0.19

```
TABLE 6 CP
 DIA.          RISE         RADIUS         ALPHA
   m             m             m            deg.
 13.900        1.733        14.802         28.00
  h1           h2            S              b            d
  mm           mm            m              mm           mm
  75           180          2.748          250          700
   y           db            qd            qb            p
  mm           mm         kN/sq.m       kN/sq.m          m
 -308           0           2.01          10.27        14.469
   A           qD            ft            fb           Vol.
 sq.m        kN/sq.m       N/sq.mm       N/sq.mm        cu.m
 161          2.583        -3.259        -2.647         17.6
   s           Nt            Mf            Nf            Qf
   m          kN/m         kNm/m          kN/m          kN/m
 0.00        -382.1        20.49         -61.40        32.65
 0.45        -356.5         8.40         -42.93        21.18
 0.90        -271.4         1.13         -25.35        11.57
 1.35        -176.5        -2.43         -11.37         4.77
 1.80         -96.6        -3.57          -1.76         0.68
 2.25         -40.1        -3.35           3.83        -1.34
 2.70          -6.0        -2.57           6.29        -1.99
 3.15          10.9        -1.68           6.63        -1.88
 3.60          16.4        -0.92           5.75        -1.44
 4.50          12.0        -0.06           2.79        -0.52
 5.40           4.4         0.15           0.30        -0.04
```

TABLE 7 CP

DIA.	RISE	RADIUS	ALPHA
m	m	m	deg.
14.300	1.783	15.228	28.00

h1	h2	S	b	d
mm	mm	m	mm	mm
75	180	2.787	250	700

y	db	qd	qb	p
mm	mm	kN/sq.m	kN/sq.m	m
-308	0	2.01	9.70	14.886

A	qD	ft	fb	Vol.
sq.m	kN/sq.m	N/sq.mm	N/sq.mm	cu.m
171	2.573	-3.196	-2.658	18.6

s	Nt	Mf	Nf	Qf
m	kN/m	kNm/m	kN/m	kN/m
0.00	-384.1	20.21	-60.23	32.03
0.45	-357.1	8.34	-42.18	20.86
0.90	-272.8	1.16	-25.06	11.48
1.35	-178.8	-2.40	-11.41	4.82
1.80	-99.3	-3.57	-1.97	0.77
2.25	-42.4	-3.39	3.57	-1.27
2.70	-7.5	-2.63	6.08	-1.96
3.15	10.2	-1.75	6.50	-1.88
3.60	16.3	-0.99	5.70	-1.47
4.50	12.5	-0.09	2.86	-0.56
5.40	4.8	0.15	0.42	-0.06

TABLE 8 CP

DIA.	RISE	RADIUS	ALPHA
m	m	m	deg.
14.800	1.845	15.761	28.00

h1	h2	S	b	d
mm	mm	m	mm	mm
75	180	2.835	250	700

y	db	qd	qb	p
mm	mm	kN/sq.m	kN/sq.m	m
-308	0	2.01	9.06	15.406

A	qD	ft	fb	Vol.
sq.m	kN/sq.m	N/sq.mm	N/sq.mm	cu.m
183	2.561	-3.121	-2.672	19.8

s	Nt	Mf	Nf	Qf
m	kN/m	kNm/m	kN/m	kN/m
0.00	-386.4	19.87	-58.84	31.29
0.45	-357.8	8.26	-41.27	20.46
0.90	-274.3	1.19	-24.68	11.37
1.35	-181.5	-2.36	-11.43	4.88
1.80	-102.4	-3.57	-2.22	0.87
2.25	-45.2	-3.44	3.27	-1.18
2.70	-9.5	-2.70	5.82	-1.91
3.15	9.2	-1.83	6.34	-1.88
3.60	16.1	-1.06	5.63	-1.50
4.50	13.1	-0.13	2.93	-0.60
5.40	5.4	0.15	0.56	-0.08

TABLE 9 CP

DIA.	RISE	RADIUS	ALPHA	
m	m	m	deg.	
15.200	1.895	16.186	28.00	
h1	h2	S	b	d
mm	mm	m	mm	mm
75	180	2.873	250	700
y	db	qd	qb	p
mm	mm	kN/sq.m	kN/sq.m	m
-308	0	2.00	8.59	15.822
A	qD	ft	fb	Vol.
sq.m	kN/sq.m	N/sq.mm	N/sq.mm	cu.m
193	2.552	-3.064	-2.683	20.8
s	Nt	Mf	Nf	Qf
m	kN/m	kNm/m	kN/m	kN/m
0.00	-388.1	19.61	-57.76	30.72
0.45	-358.2	8.19	-40.57	20.15
0.90	-275.5	1.21	-24.39	11.28
1.35	-183.5	-2.33	-11.44	4.91
1.80	-104.7	-3.57	-2.39	0.95
2.25	-47.3	-3.47	3.05	-1.11
2.70	-11.0	-2.76	5.63	-1.88
3.15	8.3	-1.90	6.21	-1.88
3.60	15.9	-1.12	5.58	-1.52
4.50	13.6	-0.16	2.99	-0.64
5.85	3.0	0.16	-0.14	0.02

TABLE 10 CP

DIA.	RISE	RADIUS	ALPHA
m	m	m	deg.
15.700	1.957	16.719	28.00

h1	h2	S	b	d
mm	mm	m	mm	mm
75	180	2.920	250	700

y	db	qd	qb	p
mm	mm	kN/sq.m	kN/sq.m	m
-308	0	2.00	8.05	16.343

A	qD	ft	fb	Vol.
sq.m	kN/sq.m	N/sq.mm	N/sq.mm	cu.m
206	2.541	-2.996	-2.695	22.1

s	Nt	Mf	Nf	Qf
m	kN/m	kNm/m	kN/m	kN/m
0.00	-390.1	19.28	-56.46	30.03
0.45	-358.8	8.10	-39.72	19.77
0.90	-276.8	1.23	-24.02	11.16
1.35	-185.9	-2.30	-11.44	4.94
1.80	-107.5	-3.57	-2.59	1.04
2.25	-49.9	-3.51	2.79	-1.03
2.70	-12.9	-2.83	5.40	-1.83
3.15	7.3	-1.97	6.06	-1.88
3.60	15.6	-1.20	5.52	-1.55
4.05	16.7	-0.60	4.37	-1.10
4.95	10.2	0.03	1.81	-0.35
5.85	3.5	0.16	-0.02	0.00

282

TABLE 11 CP

DIA.	RISE	RADIUS	ALPHA
m	m	m	deg.
16.100	2.007	17.145	28.00

h1	h2	S	b	d
mm	mm	m	mm	mm
75	180	2.957	250	700

y	db	qd	qb	p
mm	mm	kN/sq.m	kN/sq.m	m
-308	0	2.00	7.65	16.759

A	qD	ft	fb	Vol.
sq.m	kN/sq.m	N/sq.mm	N/sq.mm	cu.m
216	2.533	-2.944	-2.705	23.2

s	Nt	Mf	Nf	Qf
m	kN/m	kNm/m	kN/m	kN/m
0.00	-391.6	19.02	-55.47	29.50
0.45	-359.1	8.03	-39.07	19.48
0.90	-277.8	1.24	-23.73	11.06
1.35	-187.7	-2.28	-11.42	4.97
1.80	-109.7	-3.57	-2.73	1.10
2.25	-51.9	-3.54	2.60	-0.97
2.70	-14.4	-2.88	5.23	-1.80
3.15	6.4	-2.03	5.94	-1.87
3.60	15.3	-1.25	5.46	-1.56
4.05	16.8	-0.65	4.38	-1.13
4.95	10.7	0.01	1.88	-0.38
5.85	3.8	0.16	0.07	-0.01

TABLE 12 CP

DIA.	RISE	RADIUS	ALPHA
m	m	m	deg.
16.600	2.070	17.677	28.00

h1	h2	S	b	d
mm	mm	m	mm	mm
75	180	3.003	250	700

y	db	qd	qb	p
mm	mm	kN/sq.m	kN/sq.m	m
-308	0	2.00	7.20	17.280

A	qD	ft	fb	Vol.
sq.m	kN/sq.m	N/sq.mm	N/sq.mm	cu.m
230	2.523	-2.883	-2.715	24.6

s	Nt	Mf	Nf	Qf
m	kN/m	kNm/m	kN/m	kN/m
0.00	-393.4	18.69	-54.27	28.86
0.45	-359.5	7.93	-38.27	19.12
0.90	-278.9	1.25	-23.37	10.94
1.35	-189.7	-2.25	-11.40	4.99
1.80	-112.2	-3.56	-2.90	1.18
2.25	-54.3	-3.57	2.37	-0.90
2.70	-16.2	-2.94	5.02	-1.75
3.15	5.3	-2.11	5.80	-1.86
3.60	14.9	-1.32	5.40	-1.58
4.05	17.0	-0.70	4.38	-1.16
4.95	11.3	-0.02	1.96	-0.42
6.30	2.0	0.15	-0.37	0.05

TABLE 13 CP

DIA.		RISE		RADIUS		ALPHA	
m		m		m		deg.	
17.100		2.132		18.210		28.00	
h1	h2		S		b		d
mm	mm		m		mm		mm
75	180		3.047		250		700
y	db		qd		qb		p
mm	mm		kN/sq.m		kN/sq.m		m
-308	0		1.99		6.79		17.800
A	qD		ft		fb		Vol.
sq.m	kN/sq.m		N/sq.mm		N/sq.mm		cu.m
244	2.513		-2.825		-2.725		26.0

s	Nt	Mf	Nf	Qf
m	kN/m	kNm/m	kN/m	kN/m
0.00	-395.0	18.38	-53.11	28.24
0.45	-359.9	7.83	-37.51	18.77
0.90	-279.9	1.25	-23.02	10.82
1.35	-191.7	-2.23	-11.36	5.00
1.80	-114.7	-3.56	-3.04	1.25
2.25	-56.6	-3.60	2.16	-0.82
2.70	-18.0	-3.00	4.82	-1.71
3.15	4.2	-2.18	5.66	-1.85
3.60	14.5	-1.39	5.33	-1.60
4.05	17.1	-0.76	4.38	-1.19
5.40	8.1	0.10	1.05	-0.20
6.30	2.3	0.15	-0.27	0.04

TABLE 14 CP

DIA.	RISE	RADIUS	ALPHA
m	m	m	deg.
17.600	2.194	18.742	28.00

h1	h2	S	b	d
mm	mm	m	mm	mm
75	180	3.092	250	700

y	db	qd	qb	p
mm	mm	kN/sq.m	kN/sq.m	m
-308	0	1.99	6.41	18.321

A	qD	ft	fb	Vol.
sq.m	kN/sq.m	N/sq.mm	N/sq.mm	cu.m
258	2.504	-2.771	-2.734	27.4

s	Nt	Mf	Nf	Qf
m	kN/m	kNm/m	kN/m	kN/m
0.00	-396.5	18.07	-52.00	27.65
0.45	-360.2	7.73	-36.77	18.44
0.90	-280.9	1.25	-22.68	10.70
1.35	-193.5	-2.20	-11.32	5.01
1.80	-117.0	-3.56	-3.18	1.32
2.25	-58.8	-3.63	1.96	-0.76
2.70	-19.8	-3.05	4.63	-1.66
3.15	3.1	-2.24	5.52	-1.84
3.60	14.0	-1.46	5.26	-1.61
4.05	17.1	-0.82	4.37	-1.22
4.50	15.8	-0.36	3.24	-0.82
5.40	8.6	0.09	1.13	-0.23
6.75	1.0	0.13	-0.54	0.07

TABLE 15 CP

DIA.	RISE	RADIUS	ALPHA
m	m	m	deg.
18.000	2.244	19.168	28.00

h1	h2	S	b	d
mm	mm	m	mm	mm
90	200	3.334	280	800

y	db	qd	qb	p
mm	mm	kN/sq.m	kN/sq.m	m
-349	0	2.37	8.82	18.737

A	qD	ft	fb	Vol.
sq.m	kN/sq.m	N/sq.mm	N/sq.mm	cu.m
270	2.929	-2.347	-2.109	33.6

s	Nt	Mf	Nf	Qf
m	kN/m	kNm/m	kN/m	kN/m
0.00	-343.1	19.03	-49.14	26.13
0.45	-322.1	9.07	-36.28	18.21
0.90	-262.4	2.48	-23.91	11.31
1.35	-191.3	-1.34	-13.45	5.98
1.80	-125.2	-3.14	-5.48	2.28
2.25	-71.6	-3.61	-0.01	0.00
2.70	-32.7	-3.31	3.29	-1.19
3.15	-7.4	-2.65	4.87	-1.64
3.60	7.0	-1.90	5.23	-1.62
4.05	13.5	-1.22	4.81	-1.37
4.50	14.8	-0.68	3.97	-1.03
5.85	7.4	0.09	1.11	-0.21
6.75	2.6	0.15	-0.14	0.02

TABLE 16 CP

DIA.	RISE	RADIUS	ALPHA
m	m	m	deg.
18.500	2.307	19.701	28.00

h1	h2	S	b	d
mm	mm	m	mm	mm
90	200	3.380	280	800

y	db	qd	qb	p
mm	mm	kN/sq.m	kN/sq.m	m
-349	0	2.36	8.35	19.258

A	qD	ft	fb	Vol.
sq.m	kN/sq.m	N/sq.mm	N/sq.mm	cu.m
286	2.920	-2.303	-2.117	35.3

s	Nt	Mf	Nf	Qf
m	kN/m	kNm/m	kN/m	kN/m
0.00	-344.6	18.75	-48.20	25.63
0.45	-322.5	8.98	-35.62	17.91
0.90	-263.1	2.48	-23.56	11.18
1.35	-192.8	-1.32	-13.36	5.97
1.80	-127.1	-3.13	-5.56	2.33
2.25	-73.6	-3.62	-0.17	0.07
2.70	-34.5	-3.35	3.10	-1.14
3.15	-8.7	-2.71	4.71	-1.61
3.60	6.2	-1.97	5.12	-1.62
4.05	13.2	-1.29	4.76	-1.38
4.50	14.9	-0.74	3.96	-1.06
5.85	7.9	0.07	1.18	-0.23
7.20	1.3	0.14	-0.44	0.05

TABLE 17 CP

DIA.		RISE		RADIUS		ALPHA
m		m		m		deg.
19.100		2.381		20.340		28.00

h1	h2	S	b	d
mm	mm	m	mm	mm
90	200	3.435	280	800

y	db	qd	qb	p
mm	mm	kN/sq.m	kN/sq.m	m
-349	0	2.36	7.83	19.882

A	qD	ft	fb	Vol.
sq.m	kN/sq.m	N/sq.mm	N/sq.mm	cu.m
304	2.909	-2.253	-2.126	37.5

s	Nt	Mf	Nf	Qf
m	kN/m	kNm/m	kN/m	kN/m
0.00	-346.1	18.42	-47.11	25.05
0.45	-322.8	8.86	-34.85	17.56
0.90	-263.9	2.47	-23.15	11.03
1.35	-194.4	-1.29	-13.25	5.95
1.80	-129.4	-3.11	-5.65	2.39
2.25	-76.0	-3.63	-0.36	0.14
2.70	-36.6	-3.39	2.90	-1.08
3.15	-10.3	-2.77	4.54	-1.57
3.60	5.3	-2.04	5.00	-1.61
4.05	12.8	-1.36	4.70	-1.40
4.50	15.0	-0.80	3.96	-1.08
4.95	14.0	-0.39	3.04	-0.76
5.85	8.4	0.05	1.27	-0.26
7.20	1.6	0.15	-0.34	0.05

TABLE 18 CP

DIA.	RISE	RADIUS	ALPHA
m	m	m	deg.
19.700	2.456	20.979	28.00

h1	h2	S	b	d
mm	mm	m	mm	mm
90	200	3.488	280	800

y	db	qd	qb	p
mm	mm	kN/sq.m	kN/sq.m	m
-349	0	2.36	7.36	20.507

A	qD	ft	fb	Vol.
sq.m	kN/sq.m	N/sq.mm	N/sq.mm	cu.m
324	2.898	-2.206	-2.134	39.8

s	Nt	Mf	Nf	Qf
m	kN/m	kNm/m	kN/m	kN/m
0.00	-347.6	18.10	-46.06	24.50
0.45	-323.1	8.74	-34.11	17.21
0.90	-264.6	2.46	-22.75	10.87
1.35	-195.9	-1.26	-13.13	5.93
1.80	-131.5	-3.09	-5.72	2.44
2.25	-78.3	-3.65	-0.53	0.21
2.70	-38.6	-3.43	2.70	-1.02
3.15	-11.8	-2.83	4.37	-1.54
3.60	4.3	-2.12	4.88	-1.60
4.05	12.3	-1.43	4.63	-1.41
4.50	15.0	-0.86	3.95	-1.11
4.95	14.3	-0.43	3.07	-0.79
6.30	6.1	0.13	0.66	-0.13
7.65	0.7	0.13	-0.52	0.06

```
TABLE  19  CP
```

DIA.	RISE	RADIUS	ALPHA
m	m	m	deg.
20.300	2.531	21.617	28.00

h1	h2	S	b	d
mm	mm	m	mm	mm
90	200	3.541	280	800

y	db	qd	qb	p
mm	mm	kN/sq.m	kN/sq.m	m
-349	0	2.35	6.93	21.131

A	qD	ft	fb	Vol.
sq.m	kN/sq.m	N/sq.mm	N/sq.mm	cu.m
344	2.888	-2.162	-2.141	42.1

s	Nt	Mf	Nf	Qf
m	kN/m	kNm/m	kN/m	kN/m
0.00	-348.9	17.78	-45.06	23.96
0.45	-323.4	8.61	-33.41	16.88
0.90	-265.2	2.45	-22.36	10.72
1.35	-197.3	-1.24	-13.00	5.91
1.80	-133.4	-3.08	-5.77	2.48
2.25	-80.4	-3.65	-0.68	0.28
2.70	-40.6	-3.47	2.52	-0.96
3.15	-13.4	-2.89	4.20	-1.50
3.60	3.3	-2.18	4.76	-1.59
4.05	11.9	-1.50	4.57	-1.42
4.50	14.9	-0.92	3.94	-1.13
4.95	14.5	-0.48	3.10	-0.82
6.30	6.6	0.11	0.74	-0.15
7.65	0.9	0.13	-0.44	0.06

TABLE 20 CP

DIA.	RISE	RADIUS	ALPHA
m	m	m	deg.
20.900	2.606	22.256	28.00

h1	h2	S	b	d
mm	mm	m	mm	mm
90	200	3.593	280	800

y	db	qd	qb	p
mm	mm	kN/sq.m	kN/sq.m	m
-349	0	2.35	6.54	21.756

A	qD	ft	fb	Vol.
sq.m	kN/sq.m	N/sq.mm	N/sq.mm	cu.m
364	2.878	-2.120	-2.148	44.4

s	Nt	Mf	Nf	Qf
m	kN/m	kNm/m	kN/m	kN/m
0.00	-350.2	17.47	-44.10	23.45
0.45	-323.6	8.49	-32.73	16.56
0.90	-265.8	2.43	-21.98	10.57
1.35	-198.6	-1.22	-12.88	5.88
1.80	-135.3	-3.06	-5.82	2.52
2.25	-82.5	-3.66	-0.82	0.34
2.70	-42.5	-3.50	2.35	-0.91
3.15	-14.9	-2.95	4.05	-1.47
3.60	2.3	-2.25	4.65	-1.58
4.05	11.4	-1.57	4.51	-1.43
4.50	14.8	-0.98	3.92	-1.16
4.95	14.8	-0.53	3.12	-0.85
5.40	12.8	-0.21	2.28	-0.57
6.75	4.6	0.15	0.28	-0.05
8.10	0.2	0.11	-0.54	0.07

```
TABLE  21  CP
  DIA.          RISE         RADIUS        ALPHA
    m            m              m           deg.
 21.500        2.681        22.895        28.00
   h1       h2        S         b            d
   mm       mm        m         mm           mm
   90       200      3.644      280          800
    y       db       qd         qb           p
   mm       mm    kN/sq.m    kN/sq.m         m
 -349       0      2.35        6.18        22.380
    A       qD        ft         fb         Vol.
  sq.m   kN/sq.m   N/sq.mm    N/sq.mm       cu.m
  386     2.869    -2.081     -2.155        46.9
    s       Nt        Mf         Nf          Qf
    m      kN/m     kNm/m      kN/m         kN/m
 0.00    -351.3    17.17     -43.17        22.96
 0.45    -323.7     8.37     -32.07        16.26
 0.90    -266.3     2.41     -21.62        10.43
 1.35    -199.8    -1.20     -12.75         5.84
 1.80    -137.0    -3.05      -5.86         2.55
 2.25     -84.4    -3.67      -0.95         0.39
 2.70     -44.3    -3.53       2.19        -0.85
 3.15     -16.4    -3.00       3.90        -1.43
 3.60       1.3    -2.31       4.54        -1.56
 4.05      10.8    -1.63       4.45        -1.43
 4.50      14.7    -1.04       3.90        -1.17
 4.95      15.0    -0.58       3.14        -0.88
 5.40      13.1    -0.25       2.32        -0.60
 6.75       5.0     0.15       0.35        -0.07
 8.10       0.4     0.12      -0.48         0.07
```

TABLE 22 CP

DIA.	RISE	RADIUS	ALPHA
m	m	m	deg.
22.100	2.755	23.534	28.00

h1	h2	S	b	d
mm	mm	m	mm	mm
90	200	3.695	280	800

y	db	qd	qb	p
mm	mm	kN/sq.m	kN/sq.m	m
-349	0	2.34	5.85	23.005

A	qD	ft	fb	Vol.
sq.m	kN/sq.m	N/sq.mm	N/sq.mm	cu.m
408	2.860	-2.043	-2.160	49.4

s	Nt	Mf	Nf	Qf
m	kN/m	kNm/m	kN/m	kN/m
0.00	-352.3	16.87	-42.29	22.49
0.45	-323.8	8.25	-31.44	15.96
0.90	-266.7	2.38	-21.26	10.28
1.35	-200.9	-1.19	-12.62	5.81
1.80	-138.7	-3.03	-5.89	2.58
2.25	-86.3	-3.67	-1.07	0.44
2.70	-46.1	-3.56	2.04	-0.80
3.15	-17.8	-3.05	3.76	-1.39
3.60	0.2	-2.37	4.44	-1.55
4.05	10.3	-1.70	4.38	-1.44
4.50	14.6	-1.10	3.89	-1.19
4.95	15.1	-0.63	3.16	-0.90
5.40	13.5	-0.29	2.37	-0.63
6.75	5.4	0.14	0.42	-0.09
8.55	-0.1	0.10	-0.53	0.07

TABLE 23 CP

DIA.	RISE	RADIUS	ALPHA
m	m	m	deg.
22.800	2.843	24.280	28.00

h1	h2	S	b	d
mm	mm	m	mm	mm
90	200	3.753	280	800

y	db	qd	qb	p
mm	mm	kN/sq.m	kN/sq.m	m
-349	0	2.34	5.50	23.734

A	qD	ft	fb	Vol.
sq.m	kN/sq.m	N/sq.mm	N/sq.mm	cu.m
434	2.850	-2.002	-2.166	52.4

s	Nt	Mf	Nf	Qf
m	kN/m	kNm/m	kN/m	kN/m
0.00	-353.4	16.54	-41.29	21.96
0.45	-323.9	8.10	-30.74	15.62
0.90	-267.2	2.35	-20.86	10.12
1.35	-202.1	-1.18	-12.47	5.77
1.80	-140.5	-3.02	-5.92	2.60
2.25	-88.4	-3.68	-1.20	0.50
2.70	-48.1	-3.60	1.88	-0.75
3.15	-19.5	-3.10	3.61	-1.35
3.60	-0.9	-2.44	4.32	-1.53
4.05	9.6	-1.77	4.31	-1.44
4.50	14.4	-1.17	3.86	-1.21
4.95	15.2	-0.69	3.18	-0.93
5.40	13.9	-0.33	2.41	-0.66
5.85	11.4	-0.09	1.67	-0.42
7.20	3.7	0.16	0.09	-0.02
8.55	0.0	0.10	-0.48	0.07

TABLE 24 CP

DIA.	RISE	RADIUS	ALPHA
m	m	m	deg.
23.500	2.930	25.025	28.00

h1	h2	S	b	d
mm	mm	m	mm	mm
90	200	3.810	280	800

y	db	qd	qb	p
mm	mm	kN/sq.m	kN/sq.m	m
-349	0	2.34	5.17	24.462

A	qD	ft	fb	Vol.
sq.m	kN/sq.m	N/sq.mm	N/sq.mm	cu.m
461	2.840	-1.964	-2.171	55.4

s	Nt	Mf	Nf	Qf
m	kN/m	kNm/m	kN/m	kN/m
0.00	-354.3	16.21	-40.34	21.45
0.45	-323.9	7.96	-30.06	15.30
0.90	-267.6	2.32	-20.47	9.96
1.35	-203.3	-1.17	-12.32	5.72
1.80	-142.2	-3.00	-5.94	2.63
2.25	-90.4	-3.68	-1.31	0.55
2.70	-50.0	-3.63	1.73	-0.69
3.15	-21.1	-3.16	3.46	-1.31
3.60	-2.1	-2.51	4.20	-1.51
4.05	8.9	-1.84	4.24	-1.44
4.50	14.1	-1.23	3.84	-1.22
4.95	15.3	-0.74	3.19	-0.95
5.40	14.2	-0.37	2.45	-0.69
5.85	11.8	-0.12	1.73	-0.45
7.20	4.1	0.16	0.15	-0.03
9.00	-0.4	0.08	-0.48	0.06

TABLE 25 CP

DIA.		RISE		RADIUS		ALPHA
m		m		m		deg.
24.000		2.992		25.558		28.00

h1	h2	S	b	d
mm	mm	m	mm	mm
100	250	4.230	300	1000

y	db	qd	qb	p
mm	mm	kN/sq.m	kN/sq.m	m
-443	-0	2.68	6.12	24.983

A	qD	ft	fb	Vol.
sq.m	kN/sq.m	N/sq.mm	N/sq.mm	cu.m
481	3.411	-1.594	-1.447	69.5

s	Nt	Mf	Nf	Qf
m	kN/m	kNm/m	kN/m	kN/m
0.00	-291.9	19.79	-40.02	21.28
0.45	-281.3	11.37	-31.78	16.19
0.90	-247.1	5.16	-23.62	11.51
1.35	-202.0	0.90	-16.21	7.55
1.80	-154.8	-1.76	-9.92	4.41
2.25	-111.0	-3.20	-4.91	2.08
2.70	-73.5	-3.74	-1.16	0.47
3.15	-43.7	-3.70	1.45	-0.55
3.60	-21.3	-3.31	3.07	-1.11
4.05	-5.7	-2.75	3.92	-1.34
4.50	4.3	-2.14	4.17	-1.35
4.95	9.9	-1.56	4.01	-1.22
5.40	12.3	-1.05	3.58	-1.02
5.85	12.5	-0.64	3.00	-0.80
7.65	5.6	0.12	0.68	-0.13
9.00	1.3	0.15	-0.29	0.04

TABLE 26 CP

DIA.	RISE	RADIUS	ALPHA
m	m	m	deg.
24.700	3.080	26.303	28.00

h1	h2	S	b	d
mm	mm	m	mm	mm
100	250	4.291	300	1000

y	db	qd	qb	p
mm	mm	kN/sq.m	kN/sq.m	m
-443	-0	2.67	5.78	25.712

A	qD	ft	fb	Vol.
sq.m	kN/sq.m	N/sq.mm	N/sq.mm	cu.m
509	3.398	-1.561	-1.452	73.3

s	Nt	Mf	Nf	Qf
m	kN/m	kNm/m	kN/m	kN/m
0.00	-293.1	19.46	-39.18	20.84
0.45	-281.6	11.20	-31.13	15.88
0.90	-247.4	5.10	-23.19	11.33
1.35	-202.7	0.90	-15.98	7.47
1.80	-156.0	-1.74	-9.85	4.40
2.25	-112.6	-3.18	-4.96	2.11
2.70	-75.3	-3.75	-1.27	0.52
3.15	-45.3	-3.73	1.30	-0.50
3.60	-22.8	-3.36	2.93	-1.07
4.05	-6.8	-2.81	3.79	-1.32
4.50	3.5	-2.21	4.08	-1.34
4.95	9.5	-1.63	3.95	-1.22
5.40	12.2	-1.12	3.55	-1.04
5.85	12.6	-0.70	3.01	-0.82
6.30	11.7	-0.38	2.40	-0.61
7.65	6.0	0.10	0.75	-0.15
9.45	0.7	0.14	-0.40	0.05

TABLE 27 CP

DIA.	RISE	RADIUS	ALPHA
m	m	m	deg.
25.500	3.179	27.155	28.00

h1	h2	S	b	d
mm	mm	m	mm	mm
100	250	4.360	300	1000

y	db	qd	qb	p
mm	mm	kN/sq.m	kN/sq.m	m
-443	-0	2.67	5.42	26.544

A	qD	ft	fb	Vol.
sq.m	kN/sq.m	N/sq.mm	N/sq.mm	cu.m
543	3.383	-1.525	-1.458	77.8

s	Nt	Mf	Nf	Qf
m	kN/m	kNm/m	kN/m	kN/m
0.00	-294.4	19.08	-38.26	20.35
0.45	-281.8	11.02	-30.41	15.53
0.90	-247.6	5.04	-22.70	11.13
1.35	-203.5	0.91	-15.72	7.38
1.80	-157.3	-1.71	-9.77	4.39
2.25	-114.3	-3.16	-5.00	2.15
2.70	-77.2	-3.75	-1.39	0.57
3.15	-47.2	-3.76	1.14	-0.45
3.60	-24.4	-3.41	2.77	-1.03
4.05	-8.1	-2.88	3.65	-1.29
4.50	2.6	-2.28	3.97	-1.33
4.95	9.0	-1.70	3.88	-1.23
5.40	12.0	-1.19	3.52	-1.05
5.85	12.7	-0.76	3.01	-0.84
6.30	12.0	-0.43	2.43	-0.64
8.10	4.7	0.14	0.41	-0.08
9.90	0.3	0.12	-0.45	0.06

```
TABLE 28 CP
```

DIA. m		RISE m		RADIUS m		ALPHA deg.
26.200		3.267		27.900		28.00

h1 mm	h2 mm	S m	b mm	d mm
100	250	4.419	300	1000

y mm	db mm	qd kN/sq.m	qb kN/sq.m	p m
-443	-0	2.66	5.14	27.273

A sq.m	qD kN/sq.m	ft N/sq.mm	fb N/sq.mm	Vol. cu.m
573	3.371	-1.496	-1.463	81.8

s m	Nt kN/m	Mf kNm/m	Nf kN/m	Qf kN/m
0.00	-295.4	18.76	-37.49	19.94
0.45	-282.0	10.86	-29.81	15.24
0.90	-247.8	4.98	-22.30	10.95
1.35	-204.1	0.90	-15.49	7.30
1.80	-158.4	-1.70	-9.69	4.38
2.25	-115.7	-3.14	-5.03	2.17
2.70	-78.7	-3.75	-1.49	0.62
3.15	-48.7	-3.78	1.02	-0.40
3.60	-25.7	-3.45	2.64	-0.99
4.05	-9.2	-2.93	3.54	-1.27
4.50	1.8	-2.34	3.88	-1.32
4.95	8.5	-1.77	3.83	-1.23
5.40	11.8	-1.25	3.50	-1.06
5.85	12.8	-0.81	3.01	-0.86
6.30	12.2	-0.47	2.45	-0.66
6.75	10.7	-0.22	1.88	-0.47
8.10	5.0	0.14	0.47	-0.09
9.90	0.4	0.13	-0.41	0.05

TABLE 29 CP

DIA.	RISE	RADIUS	ALPHA
m	m	m	deg.
27.000	3.366	28.752	28.00

h1	h2	S	b	d
mm	mm	m	mm	mm
100	250	4.486	300	1000

y	db	qd	qb	p
mm	mm	kN/sq.m	kN/sq.m	m
-443	-0	2.66	4.84	28.106

A	qD	ft	fb	Vol.
sq.m	kN/sq.m	N/sq.mm	N/sq.mm	cu.m
608	3.358	-1.464	-1.468	86.5

s	Nt	Mf	Nf	Qf
m	kN/m	kNm/m	kN/m	kN/m
0.00	-296.5	18.41	-36.64	19.48
0.45	-282.1	10.67	-29.15	14.92
0.90	-248.0	4.92	-21.85	10.76
1.35	-204.7	0.90	-15.24	7.21
1.80	-159.5	-1.68	-9.60	4.36
2.25	-117.2	-3.13	-5.06	2.20
2.70	-80.4	-3.75	-1.59	0.66
3.15	-50.4	-3.80	0.88	-0.35
3.60	-27.3	-3.50	2.50	-0.95
4.05	-10.5	-2.99	3.41	-1.24
4.50	0.9	-2.41	3.79	-1.31
4.95	7.9	-1.84	3.76	-1.23
5.40	11.6	-1.32	3.47	-1.08
5.85	12.8	-0.88	3.01	-0.88
6.30	12.4	-0.52	2.48	-0.68
6.75	11.1	-0.26	1.93	-0.50
8.55	3.8	0.16	0.21	-0.04
10.35	0.0	0.11	-0.44	0.06

```
TABLE 30 CP
```

DIA.	RISE	RADIUS	ALPHA
m	m	m	deg.
27.800	3.466	29.604	28.00

h1	h2	S	b	d
mm	mm	m	mm	mm
100	250	4.552	300	1000

y	db	qd	qb	p
mm	mm	kN/sq.m	kN/sq.m	m
-443	-0	2.66	4.56	28.938

A	qD	ft	fb	Vol.
sq.m	kN/sq.m	N/sq.mm	N/sq.mm	cu.m
645	3.345	-1.434	-1.472	91.4

s	Nt	Mf	Nf	Qf
m	kN/m	kNm/m	kN/m	kN/m
0.00	-297.4	18.06	-35.83	19.05
0.45	-282.2	10.49	-28.52	14.62
0.90	-248.1	4.85	-21.42	10.57
1.35	-205.2	0.89	-14.99	7.12
1.80	-160.5	-1.66	-9.51	4.34
2.25	-118.6	-3.11	-5.08	2.22
2.70	-82.1	-3.75	-1.68	0.71
3.15	-52.1	-3.82	0.76	-0.30
3.60	-28.8	-3.54	2.37	-0.91
4.05	-11.7	-3.05	3.30	-1.21
4.50	0.0	-2.48	3.69	-1.29
4.95	7.3	-1.91	3.70	-1.23
5.40	11.3	-1.38	3.44	-1.09
5.85	12.8	-0.94	3.01	-0.90
6.30	12.6	-0.57	2.50	-0.71
6.75	11.4	-0.30	1.96	-0.52
7.20	9.6	-0.10	1.45	-0.36
9.00	2.7	0.17	0.01	0.00
10.80	-0.2	0.09	-0.44	0.05

TABLE 31 CP

DIA.	RISE	RADIUS	ALPHA
m	m	m	deg.
28.700	3.578	30.563	28.00

h1	h2	S	b	d
mm	mm	m	mm	mm
100	250	4.625	300	1000

y	db	qd	qb	p
mm	mm	kN/sq.m	kN/sq.m	m
-443	-0	2.65	4.28	29.875

A	qD	ft	fb	Vol.
sq.m	kN/sq.m	N/sq.mm	N/sq.mm	cu.m
687	3.331	-1.403	-1.476	97.0

s	Nt	Mf	Nf	Qf
m	kN/m	kNm/m	kN/m	kN/m
0.00	-298.4	17.68	-34.95	18.59
0.45	-282.2	10.29	-27.84	14.28
0.90	-248.2	4.77	-20.95	10.36
1.35	-205.7	0.88	-14.73	7.02
1.80	-161.6	-1.64	-9.41	4.31
2.25	-120.1	-3.10	-5.09	2.24
2.70	-83.8	-3.75	-1.77	0.75
3.15	-53.8	-3.84	0.63	-0.25
3.60	-30.4	-3.58	2.23	-0.87
4.05	-13.0	-3.11	3.17	-1.18
4.50	-1.0	-2.55	3.59	-1.28
4.95	6.7	-1.98	3.63	-1.23
5.40	10.9	-1.46	3.40	-1.10
5.85	12.7	-1.00	3.00	-0.92
6.30	12.8	-0.63	2.52	-0.73
6.75	11.7	-0.34	2.00	-0.55
7.20	10.0	-0.13	1.50	-0.39
9.00	3.1	0.17	0.07	-0.01
10.80	-0.1	0.10	-0.40	0.05

TABLE 32 CP

DIA.	RISE	RADIUS	ALPHA
m	m	m	deg.
29.500	3.678	31.415	28.00

h1	h2	S	b	d
mm	mm	m	mm	mm
100	250	4.689	300	1000

y	db	qd	qb	p
mm	mm	kN/sq.m	kN/sq.m	m
-443	-0	2.65	4.05	30.708

A	qD	ft	fb	Vol.
sq.m	kN/sq.m	N/sq.mm	N/sq.mm	cu.m
726	3.319	-1.377	-1.480	102.1

s	Nt	Mf	Nf	Qf
m	kN/m	kNm/m	kN/m	kN/m
0.00	-299.1	17.35	-34.21	18.19
0.45	-282.2	10.12	-27.26	14.00
0.90	-248.2	4.69	-20.55	10.19
1.35	-206.2	0.87	-14.50	6.93
1.80	-162.5	-1.63	-9.31	4.29
2.25	-121.4	-3.08	-5.10	2.26
2.70	-85.3	-3.75	-1.85	0.79
3.15	-55.3	-3.86	0.52	-0.21
3.60	-31.8	-3.61	2.11	-0.83
4.05	-14.2	-3.16	3.06	-1.15
4.50	-1.9	-2.61	3.50	-1.26
4.95	6.1	-2.05	3.57	-1.23
5.40	10.6	-1.52	3.36	-1.10
5.85	12.6	-1.06	2.99	-0.93
6.30	12.9	-0.68	2.53	-0.75
6.75	12.0	-0.39	2.03	-0.57
7.20	10.4	-0.17	1.55	-0.41
7.65	8.5	-0.01	1.10	-0.27
9.45	2.2	0.17	-0.08	0.02
11.25	-0.3	0.08	-0.40	0.05

TABLE 33 CP

DIA.		RISE	RADIUS	ALPHA
m		m	m	deg.
30.000		3.740	31.947	28.00
h1	h2	S	b	d
mm	mm	m	mm	mm
130	290	5.180	300	1200
y	db	qd	qb	p
mm	mm	kN/sq.m	kN/sq.m	m
-526	-0	3.40	6.62	31.229
A	qD	ft	fb	Vol.
sq.m	kN/sq.m	N/sq.mm	N/sq.mm	cu.m
751	4.169	-1.180	-1.107	132.7
s	Nt	Mf	Nf	Qf
m	kN/m	kNm/m	kN/m	kN/m
0.00	-264.4	20.47	-34.63	18.42
0.45	-257.0	13.02	-28.65	14.72
0.90	-234.1	7.19	-22.67	11.25
1.35	-202.5	2.84	-17.06	8.17
1.80	-167.4	-0.23	-12.05	5.56
2.25	-132.5	-2.24	-7.76	3.45
2.70	-100.2	-3.41	-4.24	1.82
3.15	-71.8	-3.94	-1.48	0.61
3.60	-48.0	-4.01	0.58	-0.23
4.05	-29.0	-3.78	2.02	-0.77
4.50	-14.4	-3.36	2.94	-1.07
4.95	-3.7	-2.84	3.43	-1.19
5.40	3.6	-2.30	3.58	-1.19
5.85	8.1	-1.79	3.49	-1.10
6.30	10.5	-1.32	3.22	-0.97
6.75	11.4	-0.92	2.85	-0.81
7.20	11.2	-0.59	2.42	-0.65
7.65	10.2	-0.33	1.96	-0.50
9.45	4.5	0.14	0.43	-0.08
11.70	0.2	0.13	-0.42	0.05

TABLE 34 CP

DIA.	RISE	RADIUS	ALPHA
m	m	m	deg.
30.900	3.853	32.905	28.00

h1	h2	S	b	d
mm	mm	m	mm	mm
130	290	5.257	300	1200

y	db	qd	qb	p
mm	mm	kN/sq.m	kN/sq.m	m
-526	-0	3.39	6.24	32.165

A	qD	ft	fb	Vol.
sq.m	kN/sq.m	N/sq.mm	N/sq.mm	cu.m
797	4.155	-1.153	-1.112	140.3

s	Nt	Mf	Nf	Qf
m	kN/m	kNm/m	kN/m	kN/m
0.00	-265.3	20.08	-33.85	18.00
0.45	-257.1	12.80	-28.01	14.41
0.90	-234.1	7.08	-22.19	11.03
1.35	-202.7	2.81	-16.74	8.04
1.80	-168.1	-0.22	-11.87	5.50
2.25	-133.6	-2.22	-7.70	3.44
2.70	-101.5	-3.39	-4.27	1.84
3.15	-73.3	-3.94	-1.57	0.65
3.60	-49.5	-4.03	0.46	-0.18
4.05	-30.4	-3.81	1.89	-0.72
4.50	-15.6	-3.41	2.82	-1.03
4.95	-4.7	-2.91	3.32	-1.17
5.40	2.8	-2.38	3.50	-1.18
5.85	7.7	-1.86	3.43	-1.10
6.30	10.3	-1.39	3.19	-0.98
6.75	11.4	-0.98	2.84	-0.83
7.20	11.3	-0.65	2.43	-0.67
7.65	10.5	-0.38	1.99	-0.52
9.90	3.5	0.16	0.23	-0.04
11.70	0.3	0.13	-0.37	0.05

TABLE 35 CP

DIA.	RISE	RADIUS	ALPHA
m	m	m	deg.
31.800	3.965	33.864	28.00

h1	h2	S	b	d
mm	mm	m	mm	mm
130	290	5.333	300	1200

y	db	qd	qb	p
mm	mm	kN/sq.m	kN/sq.m	m
-526	-0	3.39	5.89	33.102

A	qD	ft	fb	Vol.
sq.m	kN/sq.m	N/sq.mm	N/sq.mm	cu.m
844	4.141	-1.129	-1.115	148.0

s	Nt	Mf	Nf	Qf
m	kN/m	kNm/m	kN/m	kN/m
0.00	-266.2	19.70	-33.10	17.60
0.45	-257.2	12.57	-27.39	14.10
0.90	-234.0	6.98	-21.74	10.83
1.35	-203.0	2.77	-16.44	7.92
1.80	-168.7	-0.22	-11.71	5.45
2.25	-134.5	-2.20	-7.64	3.44
2.70	-102.7	-3.37	-4.30	1.86
3.15	-74.7	-3.93	-1.65	0.69
3.60	-51.0	-4.04	0.35	-0.14
4.05	-31.7	-3.85	1.77	-0.69
4.50	-16.8	-3.46	2.70	-1.00
4.95	-5.7	-2.97	3.22	-1.15
5.40	2.1	-2.45	3.42	-1.17
5.85	7.2	-1.93	3.38	-1.10
6.30	10.1	-1.46	3.16	-0.99
6.75	11.3	-1.05	2.83	-0.84
7.20	11.4	-0.70	2.44	-0.69
7.65	10.7	-0.43	2.02	-0.54
8.10	9.5	-0.21	1.60	-0.41
9.90	3.9	0.16	0.28	-0.06
12.15	0.1	0.12	-0.39	0.05

TABLE 36 CP

DIA.	RISE	RADIUS	ALPHA
m	m	m	deg.
32.800	4.090	34.929	28.00

h1	h2	S	b	d
mm	mm	m	mm	mm
130	290	5.417	300	1200

y	db	qd	qb	p
mm	mm	kN/sq.m	kN/sq.m	m
-526	-0	3.39	5.54	34.143

A	qD	ft	fb	Vol.
sq.m	kN/sq.m	N/sq.mm	N/sq.mm	cu.m
898	4.126	-1.103	-1.119	157.0

s	Nt	Mf	Nf	Qf
m	kN/m	kNm/m	kN/m	kN/m
0.00	-267.1	19.29	-32.30	17.18
0.45	-257.3	12.33	-26.74	13.78
0.90	-234.0	6.86	-21.25	10.61
1.35	-203.2	2.73	-16.11	7.79
1.80	-169.3	-0.21	-11.52	5.39
2.25	-135.5	-2.18	-7.58	3.42
2.70	-104.1	-3.36	-4.32	1.88
3.15	-76.2	-3.93	-1.73	0.73
3.60	-52.5	-4.06	0.24	-0.10
4.05	-33.2	-3.89	1.65	-0.64
4.50	-18.1	-3.51	2.58	-0.97
4.95	-6.7	-3.04	3.11	-1.13
5.40	1.3	-2.52	3.33	-1.16
5.85	6.6	-2.01	3.32	-1.10
6.30	9.8	-1.53	3.12	-1.00
6.75	11.3	-1.12	2.82	-0.86
7.20	11.5	-0.76	2.45	-0.71
7.65	10.9	-0.48	2.04	-0.57
8.10	9.9	-0.25	1.64	-0.43
10.35	3.1	0.17	0.12	-0.02
12.60	-0.1	0.10	-0.39	0.05

TABLE 37 CP

DIA.	RISE	RADIUS	ALPHA
m	m	m	deg.
33.800	4.214	35.994	28.00

h1	h2	S	b	d
mm	mm	m	mm	mm
130	290	5.499	300	1200

y	db	qd	qb	p
mm	mm	kN/sq.m	kN/sq.m	m
-526	-0	3.38	5.22	35.184

A	qD	ft	fb	Vol.
sq.m	kN/sq.m	N/sq.mm	N/sq.mm	cu.m
953	4.112	-1.079	-1.122	166.1

s	Nt	Mf	Nf	Qf
m	kN/m	kNm/m	kN/m	kN/m
0.00	-267.8	18.90	-31.54	16.77
0.45	-257.3	12.10	-26.12	13.47
0.90	-233.9	6.74	-20.78	10.39
1.35	-203.3	2.69	-15.80	7.66
1.80	-169.8	-0.21	-11.34	5.32
2.25	-136.4	-2.16	-7.51	3.41
2.70	-105.3	-3.34	-4.33	1.90
3.15	-77.6	-3.93	-1.80	0.76
3.60	-54.0	-4.08	0.14	-0.06
4.05	-34.6	-3.92	1.53	-0.60
4.50	-19.3	-3.56	2.46	-0.94
4.95	-7.8	-3.10	3.01	-1.10
5.40	0.5	-2.59	3.25	-1.14
5.85	6.1	-2.08	3.25	-1.10
6.30	9.5	-1.61	3.09	-1.00
6.75	11.1	-1.19	2.81	-0.87
7.20	11.6	-0.82	2.45	-0.73
7.65	11.1	-0.53	2.07	-0.59
8.10	10.1	-0.30	1.68	-0.45
8.55	8.8	-0.12	1.30	-0.33
10.80	2.4	0.18	-0.01	0.00
13.05	-0.3	0.09	-0.38	0.05

TABLE 38 CP

DIA. m	RISE m	RADIUS m	ALPHA deg.
34.800	4.339	37.059	28.00

h1 mm	h2 mm	S m	b mm	d mm
130	290	5.579	300	1200

y mm	db mm	qd kN/sq.m	qb kN/sq.m	p m
-526	-0	3.38	4.92	36.225

A sq.m	qD kN/sq.m	ft N/sq.mm	fb N/sq.mm	Vol. cu.m
1010	4.099	-1.057	-1.125	175.5

s m	Nt kN/m	Mf kNm/m	Nf kN/m	Qf kN/m
0.00	-268.5	18.51	-30.82	16.39
0.45	-257.2	11.87	-25.53	13.18
0.90	-233.8	6.62	-20.34	10.19
1.35	-203.4	2.65	-15.50	7.53
1.80	-170.3	-0.22	-11.17	5.26
2.25	-137.3	-2.15	-7.44	3.39
2.70	-106.4	-3.33	-4.34	1.92
3.15	-78.9	-3.93	-1.86	0.79
3.60	-55.4	-4.09	0.04	-0.02
4.05	-35.9	-3.95	1.42	-0.57
4.50	-20.5	-3.61	2.35	-0.91
4.95	-8.8	-3.16	2.91	-1.08
5.40	-0.3	-2.66	3.17	-1.13
5.85	5.5	-2.15	3.20	-1.10
6.30	9.1	-1.68	3.05	-1.01
6.75	11.0	-1.25	2.79	-0.88
7.20	11.6	-0.89	2.46	-0.74
7.65	11.3	-0.58	2.09	-0.61
8.10	10.4	-0.34	1.71	-0.47
8.55	9.2	-0.15	1.34	-0.35
9.00	7.7	-0.02	1.00	-0.25
11.25	1.8	0.17	-0.10	0.02
13.50	-0.4	0.08	-0.36	0.04

TABLE 39 CP

DIA.	RISE		RADIUS	ALPHA
m	m		m	deg.
35.000	4.364		37.272	28.00
h1	h2	S	b	d
mm	mm	m	mm	mm
130	290	5.595	300	1200
y	db	qd	qb	p
mm	mm	kN/sq.m	kN/sq.m	m
-526	-0	3.38	4.87	36.433
A	qD	ft	fb	Vol.
sq.m	kN/sq.m	N/sq.mm	N/sq.mm	cu.m
1022	4.096	-1.052	-1.126	177.4
s	Nt	Mf	Nf	Qf
m	kN/m	kNm/m	kN/m	kN/m
0.00	-268.6	18.44	-30.67	16.31
0.45	-257.2	11.82	-25.41	13.12
0.90	-233.7	6.60	-20.25	10.15
1.35	-203.5	2.64	-15.44	7.50
1.80	-170.4	-0.22	-11.13	5.25
2.25	-137.5	-2.14	-7.42	3.39
2.70	-106.7	-3.33	-4.34	1.92
3.15	-79.2	-3.92	-1.87	0.80
3.60	-55.6	-4.09	0.02	-0.01
4.05	-36.2	-3.96	1.40	-0.56
4.50	-20.8	-3.62	2.33	-0.90
4.95	-9.0	-3.17	2.89	-1.08
5.40	-0.5	-2.67	3.15	-1.13
5.85	5.4	-2.17	3.18	-1.10
6.30	9.0	-1.69	3.04	-1.01
6.75	11.0	-1.27	2.79	-0.89
7.20	11.6	-0.90	2.46	-0.75
7.65	11.3	-0.59	2.09	-0.61
8.10	10.5	-0.35	1.71	-0.48
8.55	9.2	-0.16	1.35	-0.36
9.00	7.8	-0.02	1.01	-0.25
11.25	1.8	0.17	-0.09	0.02
13.50	-0.4	0.08	-0.35	0.04

TABLE 40 CP

DIA.	RISE	RADIUS	ALPHA
m	m	m	deg.
36.100	4.501	38.443	28.00

h1	h2	S	b	d
mm	mm	m	mm	mm
130	290	5.683	300	1200

y	db	qd	qb	p
mm	mm	kN/sq.m	kN/sq.m	m
-526	-0	3.37	4.57	37.578

A	qD	ft	fb	Vol.
sq.m	kN/sq.m	N/sq.mm	N/sq.mm	cu.m
1087	4.082	-1.030	-1.128	188.1

s	Nt	Mf	Nf	Qf
m	kN/m	kNm/m	kN/m	kN/m
0.00	-269.2	18.03	-29.92	15.91
0.45	-257.1	11.58	-24.79	12.81
0.90	-233.6	6.47	-19.78	9.93
1.35	-203.5	2.59	-15.12	7.37
1.80	-170.9	-0.22	-10.95	5.18
2.25	-138.3	-2.13	-7.34	3.37
2.70	-107.9	-3.31	-4.34	1.93
3.15	-80.6	-3.92	-1.93	0.83
3.60	-57.1	-4.10	-0.08	0.03
4.05	-37.6	-3.99	1.29	-0.52
4.50	-22.1	-3.67	2.22	-0.86
4.95	-10.1	-3.23	2.79	-1.05
5.40	-1.3	-2.74	3.07	-1.11
5.85	4.8	-2.24	3.12	-1.09
6.30	8.6	-1.77	3.00	-1.01
6.75	10.8	-1.34	2.77	-0.90
7.20	11.6	-0.96	2.46	-0.76
7.65	11.5	-0.65	2.11	-0.63
8.10	10.7	-0.40	1.74	-0.50
8.55	9.6	-0.20	1.39	-0.38
9.00	8.2	-0.05	1.06	-0.27
11.70	1.3	0.17	-0.16	0.03
13.95	-0.4	0.07	-0.33	0.04

Index

The several computer programs used for outputting the design tables in this book are available from the author from £99 on 5.25 inch disks in IBM PC format. Payment, from those outside the UK, needs to be made by cheque in pounds sterling drawn on an English bank.

The programs are written in Locomotive Basic on an Amstrad System Unit PC1512 HD20 (IBM compatible). These programs are available directly from the author. The programs can be modified for use by other IBM compatibles by arrangement with the author.

The programs deal with all the designs given in the tables of this book with no need for interpolation between quantities. The Disclaimer of warranty specified in the book applies to these disks also.

The disks can be obtained by contacting the author directly at the following address:

Professor C B Wilby, Flat 27, Esplanade Court, Harrogate, North Yorkshire, HG2 0LW, England, UK (Telephone: Great Britain 0423 569061)